R&D and Licensing

Building value through intellectual assets

R&D and Licensing

Building value through intellectual assets

Kieran Comerford

Chandos Publishing
Oxford • England

Chandos Publishing (Oxford) Limited
Chandos House
5 & 6 Steadys Lane
Stanton Harcourt
Oxford OX29 5RL
UK
Tel: +44 (0) 1865 884447 Fax: +44 (0) 1865 884448
Email: info@chandospublishing.com
www.chandospublishing.com

First published in Great Britain in 2007

ISBN:
978 1 84334 236 6 (paperback)
978 1 84334 237 3 (hardback)
1 84334 236 7 (paperback)
1 84334 237 5 (hardback)

British Library Cataloguing-in-Publication Data.

A catalogue record for this book is available from the British Library.

Typeset by Replika Press, India

Printed in the UK and USA.

Contents

Acknowledgements

A number of people have given valuable assistance in the preparation of this book. My thanks to Dr Breffni Tomlin from University College Dublin and Michael Hennessy from the University of Limerick who helped me produce the earlier textbook version entitled *Managing Technology and Intellectual Assets*. My thanks also to Sean McNulty for his input on market-led innovation and project management, to Pat Branigan for his input on creativity and to David Byrne, Pauric Cregg, Donnchadh Cullinan and Maria de Aurrecoechea who provided an input to the case examples.

List of figures

List of tables

About the author

Kieran Comerford BE, MBA, qualified as an electrical engineer and worked for Philips and Westinghouse before joining the Institute for Industrial Research and Standards and working in the area of commercialising new technology. After receiving his MBA he set up a successful consultancy business and has been associated with a number of successful companies. He is well known as a lecturer on R&D management and intellectual property management and is a member of the Licensing Executives Society.

Also by the same author:

How to Find and License New Products

The Christian's Dilemma: A guide to the new spirituality

Managing Technology and Intellectual Assets

Website: http://www.kcomerford.com
The author may be contacted by email: Kieran@kcomerford.com

Introduction

Technology-based firms are now the leading drivers of economic growth in sectors ranging from manufacturing to financial services. Companies are increasingly being valued by the markets on the basis of their levels of innovation and their new product offerings. The value of a technology-based company is now determined by its intangible or its intellectual assets as they are now called. These represent the brainpower of the company, the creativity of its employees and the systems it has put in place to develop, protect and exploit these aspects of its business. However, management systems have failed to recognise the central position of technology in many firms, splitting it across departments and divisions, between cost centres and profit centres. Technology comes in many forms and is used in many parts of a business but the principles of its management are common to all technologies. This book seeks to provide an integrated approach to the management of all technologies and it does this by recognising technology as a resource and one that is central to the business.

As technology becomes recognised as a resource in a firm as in the case of finance or human resources, it becomes clear that its management is the key to success in a modern technology-based business. Management of a resource includes the idea of stewardship – looking after and safeguarding the resource. However, technology management goes further than this as it includes the tasks of creation of the resource, acquisition of additions to, and strengthening of the resource and finally, exploitation. This kind of exploitation is not based on using up the resource but on creating continuing added value. Thus technology management includes the organisation of innovation within the firm and the management of R&D. It also deals with how technology crosses the boundaries of the firm, both entering it and leaving it. Thus, it includes subjects such as intellectual property and licensing. It requires a blend of three kinds of skills, technical, commercial and legal and the successful technology manager spans these fields, operating with confidence in each while knowing when to call in the specialists.

Technology management is not technology specific. Its principles apply equally to all types of technology. It treats technology as a core element in the business and recognises the need for multiskilled individuals and multidisciplinary teams to manage it. The management principles differ from those traditionally employed in the management of R&D because they go much further, incorporating the function of business development and defining R&D as a subset of a wider subject called technology acquisition. Similarly, production is not always the outcome. Instead, it is part of the wider subject of commercial exploitation, which may take many forms.

The first part of the book deals with the subject of innovation. It is essential to have an understanding of how to introduce new products and processes successfully before going into the details of how the process is managed and the specific techniques that must be mastered and applied. The treatment follows the innovation process from idea stage right through to commercialisation. It goes on to deal with the subject of 'open innovation' a term coined by Henry Chesbrough of Harvard Business School to describe the new phenomenon whereby many companies now seek many of their innovations from outside the firm and often commercialise their technology through outside organisations. A key principle developed in this book is how open innovation can be applied in practical terms on a day-to day basis. Thus, the book treats R&D as just one method of technology acquisition, others being the purchase or licensing of technology and the use of public domain or unprotected technology developed by others.

The appropriate technology acquisition strategies for a company result from the consideration of its business strategy. Both market-led and technology-led business strategies are used to focus the new product effort to stay within clearly defined strategic objectives. Multidisciplinary teams with technical and marketing skills are used to select hunting grounds for new business opportunities. Increasingly companies are finding that traditional market research is ineffective in assessing the demand for new products. New techniques in market-led innovation are being used to predict customer needs rather than wants and to develop new products and services that effectively meet these needs.

Creativity is a source of new product ideas and of technical solutions and both of these aspects are explained and related to the traditional product development process. Companies have used Stage-Gate™ for many years as a successful technique for project management in R&D. While Stage-Gate™ is covered, the book concerns itself more with recent thinking on simplifying and improving the effectiveness of R&D project management. This is done by introducing market-led inputs to the project management process.

As stated above, R&D is only one source for the acquisition of technology. The principles of open innovation show how technology can be acquired externally. In assessing external technology, it is essential to know if it comes with intellectual property protecting it and so a significant section of this book is devoted to giving the reader a detailed and practical knowledge of this subject. This is then applied to assessing external technologies and to obtaining the optimum protection for those developed in-house.

With an understanding of intellectual property it is possible to deal effectively with the subject of intellectual assets and how these can be identified, managed and leveraged for maximum company value. Many new approaches pioneered by Dow Chemical in the 1990s have now been refined and are applied in leading technology-based companies to create strong technology positions and significant investor value for the future.

Very few managers in technology-based firms have an understanding of legal issues.

These subjects can often by clouded in jargon and inaccessible to the non-legal reader. The approach pursued in this book is that of a user of legal services rather than the provider. Contract law is explained in simple terms and the many legal contracts and documents that a technology manager may come across are described and summarised so that the reader will know when they are required and when to call in the legal professionals.

Firms can acquire technology by licensing-in and can commercialise their technology by licensing-out. However, when technology is not adequately protected, the licensing process can break down if the potential licensee sees a way to acquire the technology without paying for it. Thus the treatment of intellectual property in preceding chapters provides the foundation for engaging in the evaluation and negotiation of licences and in technology transfer. The subject of royalty rates and other licence payments is demystified and clear, simple ground rules and criteria for licence deals are provided.

Finally, the book gives some insights into the successful commercial exploitation of new technology. This includes the commercialisation of R&D projects by manufacturing or by licensing. Marketing strategies for intellectual property-based products and services lead to new business models and new approaches for commercialisation. By considering technology in the form of intellectual assets, it is shown how the traditional approach to R&D is complemented by a detailed knowledge of intellectual property and licensing, thus providing an integrated approach to the management of technology.

1

Successful innovation

Innovation is about doing something new. In the management of technology, innovation is the key to progress and competitive advantage. Companies make changes to keep ahead and maintain or increase profits, but these changes are not always successful. Many studies have been done to identify the key success factors in the development of new products and processes and to point to pitfalls that should be avoided. Lessons learned from these studies can be applied in the management of technology and organisational change at all levels.

The word innovation is derived from the Latin word *novus* meaning 'new', so innovation is about doing something new. However, doing something new may not necessarily be productive so a more comprehensive definition is needed. One such definition states that innovation is 'the successful management of change.' There can be different types of innovation depending on where the change is sought or made in an organisation. Innovation in relation to how people behave in organisations and how organisations behave and are organised is a general approach to innovation which then leads to innovation-based policies in specific areas of a company. Marketing innovation is about finding new ways to market a company's products and services or the creation of new ideas for products and services. This chapter focuses on technological innovation which is described as: 'The use of knowledge for creation or application of new technology in connection with products or processes.'[1]

Thus technological innovation is generally applied in a company, usually a manufacturing company, or in its development department. The word 'creation' implies that the company develops the new technology itself, but as we will see later, companies can also acquire technology externally and this is also part of the innovation process. A search for new ideas may be part of marketing innovation, but it may also involve applying new technical ideas and so be part of technological innovation. In innovation research the terms 'Market Pull' and 'Technology Push' are used to describe the starting point for a new idea. Innovation may come from a need identified in the market or from the application of a new technical principle.

Innovation is not the same as invention. Invention, from the Latin *invenire*, which means 'to find' is much more about the process of discovery or creation of new technical principles. William Kingston[2] sees the innovator as being at the centre of a scale he describes as:

artist – inventor – innovator – entrepreneur – trader.

The artist's side of the scale is most concerned with internal and personal activities that are not commercially driven. As you move across the scale, the activities become more externally focused, until the trader on the right has no creative input to what he sells but is purely driven by commercial factors. The innovator spans the gap between the inventor and the entrepreneur, combining a technical and a commercial focus. The innovator does not have to be the creator of new ideas, and neither does he or she have to be the entrepreneur. The innovator's role is more of a facilitator but can combine elements of all of the other roles.

The innovation process

The process of introducing new products and processes in a company is called 'the innovation process' and it consists of four distinct stages:

idea – evaluation – development – market.

The idea stage consists in identifying new ideas for products and processes and can make use of a number of techniques ranging from creativity to product and technology searching, which will be described in later chapters. Once a company has identified ideas, the next step is to evaluate them to decide whether they are likely to be successful before they go on to the development stage.

The development stage consists in identifying the tasks involved in getting from an idea to a viable product or process, and managing the successful carrying out of these tasks. Finally, the marketing or commercialisation stage involves the introduction of the new product to the market, or commercial exploitation of a new process by launching new or improved products manufactured by the process.

The stages are interrelated because the idea stage draws on both technical and market knowledge as does the evaluation stage. There is no point trying to develop new products if the technology is beyond the company's capability, or if the product will be too expensive for the market. Understanding the product specification, a description of the desired outcome of the development work, is key to the successful management of the development stage. Clearly, the product developed must be attractive to the market so that the commercialisation stage can be successful.

The product development process

The innovation process when applied specifically to product development becomes the product development process. This gives rise to more detailed stages as shown in Figure 1.1.

Figure 1.1 **Stages in the product development process**

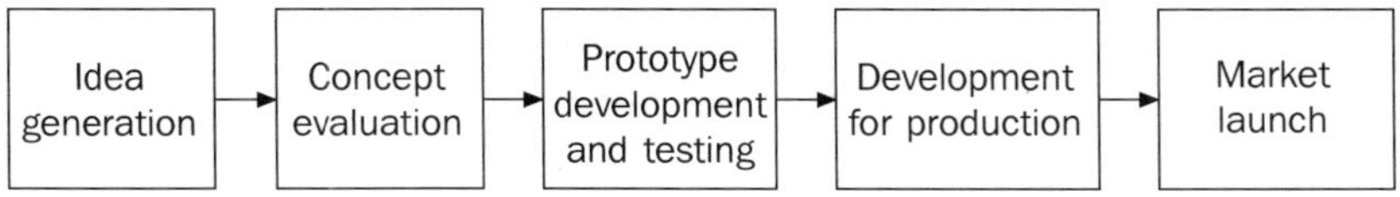

Ideas are generated from a number of sources both internal and external. These become product concepts, which are then evaluated. The promising product concepts are then prototyped, or made into working models. These are tested and the most promising approach goes forward for development. The development stage takes account of the optimum production process so that the product can be produced economically. This is called development for production. For an engineering product it could involve the manufacture of tooling for injection moulding.

The product development process is slightly different in the case of chemical and food products. Here, the product is fully developed in the laboratory and can then be tested as if it was the final product. Production development involves scaling up the laboratory process so that the product can be manufactured in quantities, as shown in Figure 1.2.

Why innovate?

Companies that do not innovate watch from the sidelines as their innovative competitors take market share from them. Robert Cooper in

Figure 1.2 Development of engineering vs. chemical and food products

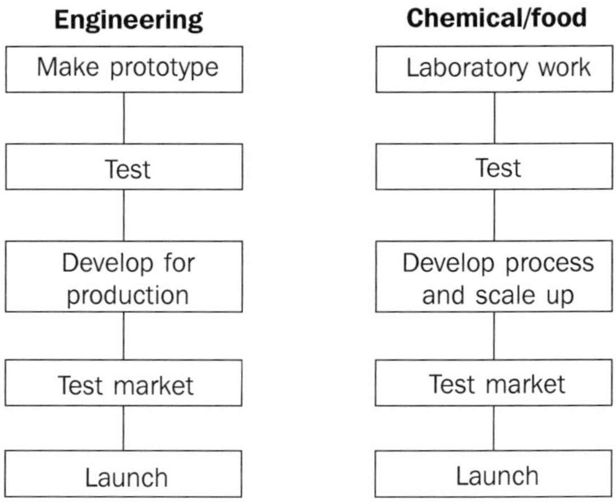

his book *Winning at New Products* describes the activities of technology-based companies as 'new product warfare' with many winners and losers.[3] He reports that 40 per cent of companies' sales are made up of products that have been introduced in the last 5 years and the figure is rising. He predicts that new products will soon account for 50 per cent of a firm's profits. The reasons for this are the ever-increasing expectations of investors. Stock markets rate innovative companies as good investments. Cooper goes on to list four key innovation drivers:

1. *Technology advances.* The increasing speed of technological development is giving rise to undreamt-of product possibilities.
2. *Changing customer needs.* Customers are spoilt for choice and they know it. They have come to expect new products that represent significant improvements and will change their buying patterns to fulfil their desires.
3. *Shortening product life cycles.* Products no longer have a life of 5–10 years. In the electronics industry some products are updated every 6 months.
4. *Increased world competition.* National markets no longer satisfy the larger corporations. The advance of globalisation leads to the manufacture of standardised products in huge volumes for a world market.

History of innovation theory: success and failure

In the nineteenth century, the number of new products arriving on the market was small, and their novelty value was considerable. Success for new inventions was almost assured. Ralph Waldo Emerson is reputed to have made the following statement in one of his lectures: 'If a man write a better book, preach a better sermon, or make a better mouse-trap than his neighbour, tho' he build his house in the woods, the world will make a beaten path to his door.'

This was in the 1800s, long before product development or marketing were recognised as disciplines. Later, as competition between companies intensified, success was not so assured and it became clear that making a better product was not the only factor leading to success.

Early innovation research derives from attempts to identify and understand the key factors leading to success and failure in new product development and introduction. After the Second World War there was an increase in economic development fuelled in no small measure by technological developments spurred on by wartime defence research. The development of radar led to significant improvements in the safety of navigation both at sea and in the air. Many completely new products became available based on the invention of the transistor by William Shockley and his colleagues Bardeen and Brattain in 1947. Economists began to advise that significant economic growth could result from the stimulation of R&D, and as a result a number of countries initiated government-funded studies to help establish government policy in fostering innovation.

A number of studies were carried out in Britain and the USA and the main findings were remarkably consistent. They all emphasised the importance attached to marketing. Good marketing skills were essential in identifying new product or service opportunities, in making sure the product or service developed would successfully meet the needs of the market and finally, in launching the product and successfully exploiting the market opportunity. They found that many innovations take place in small steps, market pull being responsible for about 75 per cent while technology push was only responsible for 25 per cent. They also found that an important method of technology transfer was personal contact. Other findings suggested that successful innovators perform their development work more efficiently but not necessarily more quickly and that they make more use of outside technology, scientific advice and information sources.

Some of the studies concerned themselves with finding out why projects failed. The results showed that commercial risks were perceived as being too great due to a lack of reliable marketing information for decision-making. Also, the heavy investment needed to obtain market share was considered too great a financial risk and uncertainty about government regulatory policies or future rulings made decisions difficult to take. These barriers to innovation can be seen to be the converse of the importance of marketing skills as a success factor.

Additionally, they found that the individual risk of being blamed for failure meant that safer options were selected. This showed a lack of understanding by management of the risks inherent in innovation and the need to quantify these risks and to seek ways of managing innovation successfully so that better success rates are consistently achieved. A significant contribution to understanding risk and success rates was made by Cooper who carried out his first studies on success and failure in new product introduction in the late 1970s.

Cooper pointed out that studying successes may show factors that they have in common, but does not indicate that these factors are related to success. Successful Japanese companies start the day with exercises and singing the company song, he says, but so do unsuccessful ones. If you studied only the successful ones you might come to the conclusion that singing the company song was a factor contributing to success. Statistics can show correlations, but correlation and causation are not the same thing.

Cooper's original project NewProd™ was a study of both successful and unsuccessful projects and analysed 102 successes and 93 failures. The study showed three important factors that distinguished successes from failures. Successful projects had:

- A unique differentiated product in the eyes of the customer.
- A product that accurately meets the needs of the market.
- Synergy and competence in technological tasks.

Secondary factors revealed by the study were:

- Marketing and managerial synergy
- Product gave savings in use.

- Strong marketing function.
- Weak competitors.

Cooper went on to design a computer-based questionnaire and scoring system for predicting the success or failure of new product concepts. This has been used successfully by many companies throughout the world.

Hewlett-Packard carried out an internal study using the Cooper system and found a number of additional factors of importance for successful projects. In particular, the Hewlett-Packard studies emphasised the importance of an integrated product introduction strategy that was correctly aligned with corporate objectives, met market needs and was supported throughout the organisation from senior management down.

Cooper's later studies went on to further refine the success versus failure criteria. In summarising these studies, a number of key success factors stand out. The most important factor is a unique product that offers a specific benefit to the user. This benefit is clearly defined and is shown to be clearly identifiable at the early stages of the project. Technical competence is also an important factor. The skill with which the technical tasks are executed results in a more successful and higher-quality product.

Successful projects had a good fit with the firm's areas of expertise or core competencies. This finding has implications for companies entering new technological areas. Another factor was the importance of carrying out the business and technical analysis fully and competently. Many projects fail where these activities are skimped or where key stages are omitted. Other factors confirmed from previous studies are the strength of the marketing function and the support of top management.

Recent research

In more recent times, research on success and failure in innovation has focused on further methods of improving success rates and dealing with rapidly changing market conditions. The initial stages of strategy and idea generation are called the 'fuzzy front end' and the term idea generation has been shortened to 'ideation'.

The fuzzy front end refers to the fact that the initial stages of product introduction have a much higher level of uncertainty than could be tolerated in the succeeding stages. It recognises this fact and even celebrates it. The fuzzy front end is where people are encouraged to be creative, allowed to spend more time on less structured tasks and where performance measurements are less quantitative and less immediate in terms of results.

As well as product life cycles, it has been observed that technologies also go though life cycles, although these are of longer duration than for products. Technology life cycles are described by S-curves and these can be used to predict when a technology is approaching maturity and when a replacement technology is needed. This subject will be developed in the section of the book on technology assessment.

All of the studies, and especially the earlier ones, emphasise the importance of a strong marketing function. While technology push was important in earlier times, market pull has become increasingly important leading to new technology strategies that integrate business and technology aspects. A strong marketing function in a company ensures that the areas that the company targets for development will be growth areas and will yield significant strategic advantage. It ensures that the ideas found and the products introduced will be relevant to market needs, and finally it ensures that the firm will be able to gain market acceptance for its new products.

Open innovation

Many organisations no longer do all of their R&D in-house. Increasingly, companies innovate by using a combination of R&D and externally sourced technologies. These may be acquired by licensing or sometimes by imitating competitors. Many of the larger companies which formerly had large R&D units are now slimming them down and getting many of their ideas from outside. Bell Labs and General Electric have cut back on R&D expenditure because they are not getting a good enough return and Procter and Gamble have announced a target of 50 per cent externally sourced technologies. Intel has acquired much of its technology from outside sources.

Formerly innovation was carried out in secret and in vertically integrated companies who commercialised their own technologies. However, many organisations found that much of their developments remained on the shelf because they did not fit their business model. Henry Chesbrough of Harvard Business School has described these trends in his book *Open Innovation: The New Imperative for Creating and Profiting from Technology*.[4] He says that the old 'Not Invented Here' syndrome is a thing of the past as companies realise that they cannot afford to develop everything they need, nor can they afford to ignore significant advances taking place outside their organisations. Open innovation also works at the output end of the process. Chesbrough describes how Xerox set up its Palo Alto Research Center and then found that some of its ideas were commercialised by people leaving the organisation.

Chesbrough labels the old model as 'closed innovation'. In this model companies built up large research facilities and tried to commercialise all their developments within the organisation. Inevitably many ideas were abandoned because they did not fit in with the company's current business model. Disgruntled researchers began to leave to pursue their ideas outside. With a vibrant venture capital market many set up successful companies, sometimes taking licences from Xerox or with Xerox taking a stake in the new company. Adobe and 3Com started in this way.

This new kind of thinking means that larger organisations are more open to new ideas from outside. It also means that they are more ready to open their R&D treasure chests to outsiders. Microsoft now announces its licensing opportunities through its website where they state 'Leveraging great ideas. It's the way to get ahead—and stay out front. And now, through Microsoft IP Ventures, we're giving small, VC-backed companies a whole new way to bring innovations to market faster and with less risk. It's a smarter way to partner. And the fast track to success.'[5]

The Apple iPod: an example of open innovation

The iPod was not initially conceived by Apple but by Tony Fadell, a young engineer who had worked at the Apple spin-off General Magic, at Philips Electronics and briefly at RealNetworks. Fadell had offered the idea to Philips and RealNetworks before it was accepted by Apple. Luckily Apple were open to his ideas, took him on and gave him the resources he needed to develop it. He was given a team of 35 people and Apple CEO Steve Jobs took an active interest in the project on a day to day basis.

The iPod is a good example of open innovation in action as it involves an idea from outside Apple and many external components and technologies. The MP3 player came from a company called PortalPlayer. The operating system was acquired from Pixo a company set up by former Apple employees. The microprocessor came from ARM in the

UK and the disk drive was supplied by Toshiba. The battery came from Sony and the data compression from a German university. One of the keys to the iPod's success was the employment of the British designer Jonathan Ive who created its stunning good looks.

The iPod sold 67 million units in its first 5 years and continues to achieve sales through updates with enhanced performance. The product has brought Apple into a completely new market and has achieved a 70 per cent share of the market for digital music players in the USA. Its links to the iTunes download store for digital media has given it a new business model incorporating a new aftermarket revenue stream. The spin-off for the Apple brand has been considerable.

Notes

1. Holt, K. (ed.) (1973) *Innovation*. I.I.M.T.
2. Kingston, W. (1977) *Innovation*. John Calder.
3. Cooper, R.G. (1993) *Winning at New Products*. Perseus, p. 4.
4. Chesbrough, H.W. (2003) *Open Innovation: The New Imperative for Creating and Profiting from Technology*. Harvard Business School Press.
5. See *http://www.microsoft.com/mscorp/ip/ventures/about.asp* (accessed 12/12/05).

2

Using technology to gain competitive advantage

Introduction

Organisations acquire technology in many different ways. The most obvious is by carrying out their own R&D, but this is not always the best way. Companies can also purchase or rent technology, or copy ideas, products and technologies of other companies. It is important to understand the different technology acquisition strategies and the implications of each from a technical, commercial and legal perspective. With an understanding of how to acquire technology we can then consider which technologies to acquire. These must align with the company's business objectives. This chapter introduces the basic principles of technology and technology acquisition strategies and begins consideration of the first stage of the innovation process – the idea stage.

Technology: the invisible resource

When we look at the key resources in any company we tend to think of finance, people, products and processes, and perhaps buildings and machinery. For all of these resources we provide means of stewardship or systems of management. However, there is one key resource that is invisible, and that is technology. Like any other resource, technology can be managed. It can be acquired, developed, maintained, bought and sold and leased. Why do we appear to have this technology blindness?

If you trace the history of industrial development, you can see how the different ingredients in the mix of any business have evolved, come to the fore, and then settled into their present position. Technology was key to the start of the industrial revolution but it was not called technology in those days. Instead people thought in terms of inventions. The invention of steam power made it possible to draw coal up from far under the ground, to transport goods over long distances and to drive the machines in the booming textile factories of the nineteenth century.

Finance quickly became the major factor as larger projects came into prospect that were beyond the financial reach of individual entrepreneurs. Many railways were built with the hard-earned cash of hopeful fledgling investors. Management techniques and then marketing came to the fore in the twentieth century and by the end of that century information was king.

The combined effect of sophisticated marketing and widespread access to information has produced a huge demand for products and services. This has led to an acceleration in product and process development. Product development lead times are down to 6 months in parts of the electronics industry and many product life cycles are so short that it is difficult for the consumer to keep track of model numbers or find spare parts.

However, the key driver of this surge in product and process development is not finance, marketing or information; it is technology, which has been developing steadily and waiting until the other necessary ingredients were in place for it to take on its lead role of today. Many people talk of a knowledge economy as if knowledge was the key ingredient, but knowledge without technology is not able to produce the new products and processes that companies need to maintain or improve their competitive position.

What is technology?

Technology is defined as the application of science. Science, from the Latin word *scientia*, means knowledge. Knowledge without being applied is just knowledge and brings no economic benefit. We hear about 'technology stocks', the 'technology sector' and 'technology companies', but technology is not just information technology, as many journalists would have us believe. Technology covers the application of physics, i.e. engineering, and the application of chemistry, i.e. chemical and food production. Information technology is the application of information science made possible through advances in engineering (i.e. electronics).

However, describing technology as the application of science does not give the full picture. Technology is only of value when it is commercial, when it is capable of delivering products and processes that give a business a competitive advantage. A company may have complex and expensive technology but if it has been superseded by a new development elsewhere, its value may be limited.

Importance of technology

Technology itself is now a major determinant of the success or value of a company. When companies change hands, the value of a company's assets are no longer reflected in the value of tangible assets on its balance sheet. Technology-based companies typically have a sale price over 10 times their book value. The value of a company is now in its intellectual assets which combine its technology, its intellectual property, the expertise of its workforce and its business model. Thus the acquisition of technology, this key resource, is fundamental to the growth, development and even survival of a company.

Technology acquisition strategies

There are many ways in which a company can acquire technology. These can range from internally conducted R&D to the purchasing of technology from outside the company. Companies traditionally did all their own R&D but more recently the idea of acquiring rights to outside developments, the concept of open innovation has become common. In fact, there are a number of strategies for technology acquisition open to modern technology-based companies. The approach in this chapter is to categorise the possible strategies for technology acquisition in relation to the sources of technology. The spectrum of sources from 100 per cent internally generated to 100 per cent externally generated gives rise to a number of strategic possibilities as shown in Table 2.1.

The sections that follow consider each of these strategies, what they mean, and their implications for a company, under the headings shown on the left-hand side of Table 2.1.

Table 2.1 Strategies for technology acquisition

	Source					
	Internal ⟵				⟶	External
Strategy	Invent	Improve	Modify	Copy	License	Takeover
Ownership	Yours	Shared	Borrow	Steal	Rent	Buy
Cost	High	Medium	Medium	Low	Medium	Low/medium
Risk	High	Medium	Low	Low	Low/medium	Low
Profit potential	High	Medium	Medium	Low	Medium	Medium/high
Protection	Excellent	Good	Fair	Poor	Good	Fair/good

Inventing

Inventing means coming up with something completely new. This is radical innovation, a new principle or a completely fresh approach. For example, lasers, the hovercraft and computer mouse pointer systems with their related hardware and software. Because these inventions are completely new and owe very little to previous developments, they belong wholly to the inventor or developer.

The cost of development is high because there is a lot of uncertainty in the development programme. Researchers may have to explore many blind alleys and carry out many tests and trials. Consequently, the risk is high but the profit potential if success is attained is also high. Because of the high level of novelty and inventiveness, it is usually possible to protect the technology against imitation or copying thereby increasing the profit potential, reducing the risk, and giving a higher return on the R&D investment.

Improving

This is what most companies do. It involves inventing but the invention is usually an improvement on an earlier version of the product or technology. The ownership is shared with the earlier version which may belong to the company or it may be an improvement on a competitor's product. In that case the company owns only what it has added to the 'state of the art'.

The cost is medium because one is building on an existing platform of development and it is not a radical innovation. The risk is also medium because you know that there is a market for the earlier version. The profit potential is medium but may be high if the improvement is significant or gives a significant reduction in manufacturing cost. The risk is also medium and it should be possible to protect the improvement. Of course, it is not possible to obtain protection extending back over the earlier version, but if this belongs to a competitor, the protection can stop the competitor copying the improvement.

Modifying

By modifying we mean introducing a product that is based on a competitor's product but that has been redesigned at least cosmetically so as to avoid the accusation of copying. The company only owns its redesign work and is borrowing the rest of the product idea or technology from a competitor. This is legitimate if the company does not infringe the competitor's intellectual property.

The cost is medium because redesigning can be almost as expensive as improving. When costing it is wise to include for the cost of investigating the extent of the competitors' intellectual property and obtaining legal advice. The risk is low because the company is entering an already existing market, trying to gain market share at the expense of a competitor. Because the level of innovation is low and the competitor's product is not well protected, it is unlikely that profit margins will be attractive. It may be possible to gain some limited protection for the company's own design elements, but if the product is not well differentiated from the competitor's, the protection will not be good enough to allow you to charge higher prices.

Copying

This is the classic 'Me too' product and is a straight copy of a competitor's product in so far as you can get away with it without legal problems. This type of product is common in consumer marketing. You see a new product on the supermarket shelf and 6 months later there are three or four imitations beside it. The company pursuing this strategy does not own the product or technology and could be accused of stealing it. This type of activity usually takes place in product areas where there is very little protection and the product idea is the only original part of the development. As we will see later, it is very hard to protect an idea, so the chances of being accused of stealing one are small.

The risk is low because you are merely entering an existing market. Margins will also be low in such a competitive environment. The entry of one imitator is usually followed by two or three others that forces prices down. Finally, the only element of competition is price and the product becomes a commodity. Product protection is non-existent or may be limited to the use of branding or trade marks.

Licensing

Licensing is about acquiring the rights to do something that you were not in a position to do before. This could involve the rights to operate under a patent, or the rights to make use of secret know-how or information that is not generally accessible, or it could involve rights to other forms of intellectual property. It implies some form of protection; otherwise companies would not pay for the rights. They would simply try to copy the technology.

Licensing is like renting the use of a technology. You do not own it but you have the use of it. The cost is medium as you usually have to pay for access to the technology and for the continuing use of it in the form of royalties or sales-related payments. The risk is low if the technology or product is already established on the market but can be higher if the technology or product is at an early stage and you have to finish the development work and commercialise it.

The profit potential is medium as some of the profits are always diverted to pay royalties but the protection for the product in the marketplace is good as indicated by the fact that you have decided that it is better to pay royalties than to attempt to copy it or try to design around its protection.

Takeover

It is also possible to buy technology by buying a patent or technology outright, or by buying the company that owns it. Buying technology outright can be expensive as you have to compensate the owner for the loss of future royalties or profits. Buying the company is better as most takeovers involve share swaps or a mixture of share swaps and cash, which has the effect of reducing the cash exposure of the purchaser. The seller is agreeing to be paid out of the future profits of the purchaser.

The risk is relatively low as the purchaser would not buy the company unless its technology already had a commercial value and there were significant synergies in acquiring the technology. The risk may be higher if the technology is still in development or the company being purchased is loss-making. This will be reflected in the price. There have been numerous cases where technology developing companies have simply run out of R&D cash and been sold off for less than their worth under pressure from creditors.

The profit potential is medium to high and this depends on the quality of the opportunity. The protection for the product or technology in the marketplace is usually reasonably good; otherwise people would try to copy it rather than purchasing the company.

Business and technology strategies

Before deciding on a technology acquisition strategy as outlined in the previous section, a company must have a technology strategy. A technology strategy is a strategy for using technology to achieve the business objectives of the company. Clearly, the first issue to address is that of deciding the company's business strategy.

The original definition of strategy is 'the science of planning and directing military operations'. This concept is now applied to business and as you can see from the definition, it includes the idea of planning, which is to do with arranging a series of actions so that a certain future outcome can be achieved. Business strategy starts with a definition of the company's mission or strategic intent. It asks questions such as 'what business are we in?' or 'where do we want to be in five years?' From this, the company develops a set of objectives and prepares a plan stating the steps needed to achieve the objectives.

Other questions concern what skills or competencies the organisation possesses, what is it good at and what is it not so good at. Also it is necessary to look at the business or regulatory environment in which the company is operating. It is only from this perspective that the company can plan how to move forward. This is called strategic analysis.

Strategic analysis

If you want to travel to a place you have not visited before, the obvious first step is to look up a map. However, you must know where you are now so that you can find your place on the map and see which road leads to your destination. If you don't know where you are now, how can you possibly make use of the map? Strangely, this is the position of many companies that assume a business strategy without basing it on a sound knowledge of their current position.

The simplest method of finding out where you are now is to do an audit of strengths and weaknesses of all the aspects of the company. A more comprehensive analysis includes looking at opportunities and threats. This is called a SWOT analysis, SWOT standing for:

Strengths **Weaknesses** **Opportunities** **Threats**

Strengths and weaknesses refer only to factors that are internal to the company, for example its technology or financial health. Opportunities and threats refer only to external factors, for example, if the company is financially weak, this is not a threat; it is a weakness, an internal factor. If the bank is looking for repayments, that may be a threat because it is an external factor. Table 2.2 shows a brief summary of a SWOT analysis for a small company. A full SWOT analysis would be much more detailed and more specific.

Having carried out this analysis, the company's objectives should be to eliminate its weaknesses and capitalise on its strengths so as to take advantage of the opportunities and counter any threats. Threats should be seen as opportunities in disguise. For example, the threat of new technology could be countered by taking on the licence opportunity and the environmental issues could give the company a chance to make its products more friendly to the environment.

Table 2.2 Summary of SWOT analysis for a smaller company

Strengths	Weaknesses	Opportunities	Threats
Technology	Marketing	Licence offer	New technology
Skilled workforce	Location	Low interest rates	Falling prices
Finance	Skills shortages	Eastern Europe	Far East competitor
Equipment	Planning	New distributor	Environmental issues

Relationship of business and technology strategies

Business and technology strategies are inextricably linked in a technology-based company. A company may decide to exploit a certain market (a business strategy) and develop the technology to enter the market (a technology strategy). Alternatively, a company may decide to exploit its technological competence in a certain area (a technology strategy) by using it to enter a certain market (a business strategy). Another approach is to engage in a diversification strategy. These points will be developed in the paragraphs below.

Diversification strategies

A strategic analysis may show a firm that it needs to upgrade its technology, but it does not necessarily point the way to new business opportunities for business growth. One approach to developing a strategy for business growth is the Ansoff Product/Market matrix illustrated in Figure 2.1.

This approach developed by H. Igor Ansoff in the 1960s gives a simple example to show how business growth strategies can lead to technology strategies. Ansoff's matrix gives four possibilities based on the combinations of existing and new products and existing and new markets. A company can stay with its existing products and markets or it can diversify into new products and new markets. It shows that there are a number of ways in which a company can attempt to increase the size of its business. These have varying degrees of risk. A principle called 'incremental innovation' suggests that you should move forward

Figure 2.1 **Ansoff Product/Market Matrix of diversification strategies**

Markets \ **Products**	Existing	New
Existing	Market penetration	Product expansion
New	Market expansion	Radical diversification

in small steps so that you can measure the effect of each change and you can pull back if things are not going according to plan. Following the incremental innovation principle, it is best to go one step at a time. This means moving one square horizontally or vertically. Moving diagonally, i.e. entering a new market with a new product, is described as radical diversification and is generally not recommended because of the high risk involved.

The four expansion strategy options in Ansoff's Product/Market Matrix lead to appropriate technology strategies. These are shown in Table 2.3.

Cost reduction

To stay in the same market with the same product is almost impossible without some change. Change is necessary even to stand still because competitors are always changing. The usual strategy here is one of market penetration to increase market share. This requires cost reduction so as to be able to reduce prices, so the technology strategy is related to process improvement or value engineering.

New product development

The business strategy for product expansion in the existing market is one of new product development. This involves the introduction of new technology, or new products using existing technology.

Localisation

In entering a new market it is often necessary to modify the product by localising it. This may require changes to conform with standards, or

Table 2.3 Business strategies leading to technology strategies

Business strategy	Technology strategy
Market penetration	Cost reduction, process improvement
Product expansion	New product development
Market expansion	Localisation
Diversification	Technology purchase

translating languages for packaging or instructions. Localisation is a major activity in software market expansion.

Technology purchase

In diversifying, companies frequently acquire new technology by purchasing companies or purchasing licences, thereby making a step change in their technological capability. As mentioned above, radical diversification involving entering a new market with a new product is a high-risk strategy. This is the same activity undertaken by entrepreneurs who start a new business with a new product. Because of the high risk involved, entrepreneurs are usually advised to form a team with other people of complementary skills. For example, an engineer is usually asked to team up with a marketing and a financial person.

Technology strategies are not the same as technology acquisition strategies. Technology strategies are how the company uses technology to achieve its business objectives. This can mean using the company's existing technology and this can be seen in some of the technology strategies above. For example, cost reduction may not necessarily involve any acquisition of new technology. New product development may simply be achieved by extending the product range within the existing technology, and localisation rarely involves significant change to the product itself. Technology strategies can also lead to the acquisition of new technology where this is needed to develop new or improved products or processes.

Technology-led strategies

Larger organisations have used sophisticated strategic analysis to identify what are called 'core competencies' and to play to their technological strengths. C. K. Prahalad and Gary Hamel[1] describe the strategy of building on and acquiring core competencies as one that is still unknown in most corporations. In contrast, however, they describe how NEC grew to become a major player in information technology by entering into over 100 strategic alliances to acquire access to new technology. They show how Honda used its competence in small lightweight engine design to exploit markets in motorcycles, outboard motors, cars and generators. Similarly 3M used its competence in adhesive technology to exploit markets in coatings and office products. The key characteristics of core competencies are:

- They give access to a wide variety of markets.
- They are difficult for competitors to imitate.
- They contribute to improved customer satisfaction.

Core competencies are described as 'the roots of competitiveness', which feed into the development of core products giving rise to business units whose fruits are end products.

To sustain leadership in the area of their chosen core competencies, companies seek to maximise their world manufacturing share in core products. For example, Canon became the leading supplier of laser printer engines, supplying most of its competitors. 3M developed a new product such as the Post-it® thereby increasing the use of its core product, i.e. adhesives.

A core-competence-based strategy has the following steps:

- Identify core competencies.
- Research product/market opportunities to exploit core competencies.
- Decide on market entry strategies and product specifications.
- Decide on product/technology development requirements.
- Acquire products or technology to implement plans.

Market-led strategies

The market-led or business strategy approach is based on market pull rather than technology push. Market research, which is used to identify a growing market, and then the technology is acquired to deliver the appropriate products. The approach has the following steps:

- Decide on a market to be in.
- Decide competitive position/strategy.
- Identify product strategy to support market strategy.
- Select technology to support product strategy.
- Acquire technology to implement plans.

The market-led approach starts with deciding what market to be in, then how to enter the market and with what products. The company should have some idea of the areas or hunting grounds where it might find ideas for new products that would fit with the company's business strategy. The market ultimately selected will result from the strategic objectives followed by the strategic analysis possibly supported with some research on market trends. Identifying the correct product offering for the market is not so easy and cannot simply be contracted out to market researchers. In both market-led and technology-led approaches, critical decisions on product and technical specifications will have to be made. From the technology-led approach suggestions will be made for products with enhanced functionality or new breakthrough capabilities. These technical advances are sometimes sufficient to ensure success as in the case of the iPod. However, many technology-led products do not succeed. Examples of this are the Sinclair C5, Sir Clive Sinclair's electric tricycle, which was panned by the press, and the Segway, which is still struggling for acceptance.

In the market-led approach, finding the right product offering may not be easy but there is a better chance that there will be a demand for it when it is launched. That is, if the correct market and product opportunity have been identified. Traditional market research has, in the author's experience, been ineffective in predicting what customers will actually buy as compared with what they say they will buy. People interviewed tend to give the answer that they think is expected of them. Increasingly companies are coming to rely on identifying customer needs rather than wants, i.e. their undeclared or even subconscious opinions and preferences. This is done by asking indirect questions about product attributes rather than asking customers to react to product samples or prototypes.

Akio Morita one of the founders of Sony famously disregarded market research relying instead on his own instinctive belief that there was a market for the Walkman, a tape player that retailers said would not sell because it could not record.[2] For the product launch he ordered an initial production run of 30 000 units, a brave decision as the monthly sales of the best-selling tape recorder at that time averaged only 15 000 units.

The key to successfully following either a market or technology-led business strategy is the use of multidisciplinary teams. Fully integrated teams focused on the same goals and working in close communication can avoid many of the conflicts that have traditionally occurred between technical and marketing disciplines. Problems have been caused by technical personnel straying from the product specification and launch date, adding extra functions and costs, or failing to disclose the extent of technical shortcomings. Similarly, marketing staff have been accused of overestimating demand and having unrealistic price expectations.

In consumer markets many customers buy products based on features that appeal more to their emotions. Predicting demand in the business to business market is easier than in the consumer market because the purchase decision is based on more rational criteria such as price and performance. Many companies have successfully identified market opportunities in selling to other businesses. Examples abound in energy, in the environment and in changes in government regulations.

Trenchlink Ltd: an example of market-led innovation

Trenchlink was founded by David Byrne, a supplier of diamond blades to the construction industry. As a result of observations made of his clients who were involved in cutting trenches for roadworks, he took a strategic decision to enter the market for trench covers and identified an opportunity for a new design of trench cover that would give significant benefits and savings in use.

He observed trends in the UK to control the extent to which roadworks were allowed to disrupt traffic especially at rush hour. Local authorities had appointed streetworks co-ordinators who would rent the road to the contractor, at a daily rate, for the duration of the works. If the contractor took longer than expected it would pay the penalty in increased rental charges. In areas of high traffic density the contractor was required to close any road openings before rush hour and would then have to open them the next day and dig out the filled trench before resuming work. Contractors were not permitted to cover trenches with sheets of steel because these could not be properly secured and the surfaces of steel sheets have a very low coefficient of friction.

Byrne saw the opportunity for an innovative design of trench cover plate that could be secured in place over the opening, and would be safe for traffic to pass over. The solution was a set of interlocking plates that are laid along the trench and cannot move in any direction once locked in place. The patented plates are made from cast iron and have antiskid surfaces that are long lasting. Following extensive testing the plates were specified by National Grid Transco for use in the replacement of the UK gas grid, a project expected to cost £30bn and run for 10 years. The plates are now used by 5000 road crews for all road openings in dense traffic areas.

Notes

1. Prahalad, C.K. and Hamel, G. (1990) The core competence of the corporation. *Harvard Business Review* 68: 79–91.
2. See *http://www.sony.net/Fun/SH/1-18/h2.html* (accessed 15/12/05).

3

Technology acquisition

Introduction

We have seen in the previous chapter that a company can acquire technology in different ways. This chapter describes a wide range of scanning techniques that companies use to look for new technologies and products and emphasises the importance of modern information systems. The focus in this chapter is on acquiring ideas or new technology from outside the company. Chapter 4 will deal with ideas internally generated.

Methods of technology acquisition

Methods of technology acquisition can start a long way ahead of any commercial objectives. Many companies, research organisations and universities carry out basic research. This type of research is not hampered by having to meet commercial criteria and may be only generally guided by business and technology strategies. The emphasis is on acquiring knowledge and then finding out if the knowledge acquired can be put to any commercial use.

The 3M Post-it® came into being because a researcher had developed an adhesive that was tacky but not sticky. For a long time no applications could be found for the product until the idea of the Post-it® came up.

The process from research to commercialisation can take 7 years or more, as Figure 3.1 shows. This figure illustrates the technology development pipeline.

Figure 3.1 The complete technology development process

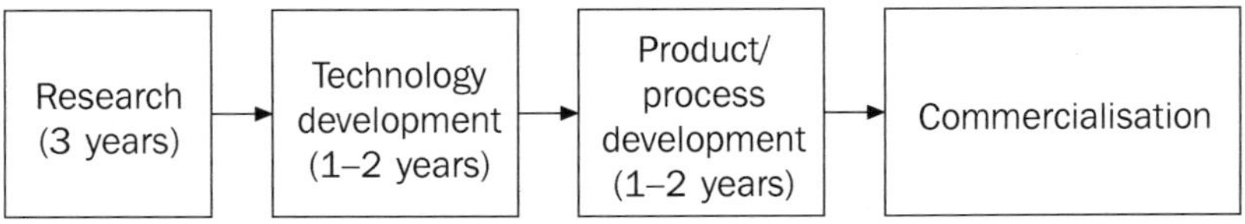

Many organisations use technology forecasting techniques to identify possible areas for developing new technologies or predicting market requirements. These are sometimes called arenas or hunting grounds. The output of these studies is used as an input to the business and technology strategy stages.

When a company has decided on its business and technology strategies, it can then select a technology acquisition strategy as described in Chapter 2. In this chapter, we look in more detail at the selection of such a strategy, especially at the question of whether technology should be sourced externally. This chapter also describes a detailed technique for external technology searching.

Why research & development?

Many companies automatically start into an R&D programme without considering the other technology acquisition strategies. This is usually the result of giving insufficient time or attention to business or technology strategy formulation. R&D, the technology acquisition strategy of inventing or improving, is a high/medium-risk strategy. Considerable financial resources must be committed long before any revenues will be generated. Then, as Cooper[1] reports, 46 per cent of all expenditure on R&D is wasted on failed projects. Only one on seven product concepts and one in four development projects succeed. Also, experience has shown that most projects overrun their original budgets by 1.5–3 times, depending on the level of innovation. So you can start an R&D project that will take twice as long as you think, cost twice as much money, and still only have a 25 per cent chance of success! If it does generate revenues, they will be 2 years in coming and you have to survive financially meanwhile.

Alternative technology acquisition strategies are worth considering, in the light of these realities. Licensing-in of an established technology

Table 3.1 R&D versus licensing

R&D	Licensing
Pay all costs upfront	Smaller upfront payment
Long time to market	Shorter time to market
Cost overruns (1.5–3 times)	Predictable costs
Might not be successful	Product proven in another market
No technical assistance	Technical assistance form licensor
No royalties payable	Royalties paid out of cashflow

from a company in another part of the world can be attractive if you can find the right technology and partner. It is not easy to do this, but it is certainly worth the effort. It is now relatively easy to search through patent databases and other sources on the Internet to look for suitable licensors. Purchasing technology-developing companies can also be a good strategy, again if the right prospect can be found. Both of these strategies can give a company access to new technology within months, at a fraction of the cost of the equivalent R&D investment. By following this strategy, a company can reduce the technical and commercial risks, and the chances of success are high, much higher than 25 per cent. Here, we are talking about a project that is finished and commercialised and not about acquiring unproven technology.

The other low-risk strategies, such as modifying or copying, generally give low rewards. However, as you will see, a good knowledge of intellectual property can be a distinct benefit in taking advantage of loopholes left by competitors in the intellectual property protection of their products and technology.

A further benefit in having a broader technology acquisition strategy is that it opens the company to a much wider range of ideas and encourages a proper evaluation of alternatives before embarking on an R&D programme. Use of modern on-line information systems can give the company a much better picture of the relationship of their proposed development to existing products and technology. Frequently, the final decision is to follow the R&D route, but the decision is taken from a much more informed perspective. The specification of the product or process developed is much better defined and much more likely to achieve commercial success.

Technology forecasting

Strategic analysis leads to the identification of strategic business opportunities. This can lead to the identification of gaps in the company's technology, or to unexploited market opportunities. A significant part of any strategic market analysis is the anticipation of future market trends and customer demands. As we saw in Chapter 2, the best approach is to integrate both business and technology into the strategic analysis. A technique called technology forecasting attempts to combine the prediction of market demand and technological capability.

Technology forecasting is defined as 'the description or prediction of some foreseeable technological innovation, specific scientific refinement or likely scientific discovery, that promises to serve some useful function, with some indication of the most probable time of occurrence.'[2] The emphasis here is mainly on forecasting technological developments, but the inclusion of the words 'that promises to serve some useful function', ensures that the technique has market relevance. Many attempts at technology forecasting have been very wide of the mark, as the following famous forecasting mistakes show:

- Total market for PCs according to IBM in 1981 was 241 683.
- Electric cars were expected to be in widespread use by 1996.
- The Videophone was expected to be in widespread use by 1980s.
- Limitless supplies of nuclear power were expected by1966.

Technology forecasting methods

The methods employed in technology forecasting range from simple to complex and involve procedures that rely on judgement or that may be more mathematically based. Figure 3.2 lists a number of methods and their relationship on a scale of simple to complex and judgemental to mathematical.

The method most often used is the Delphi technique. The other techniques are described in *Managing Technological Innovation* by B. C. Twiss.[3] Delphi is popular because it is neither too complex nor too mathematical. It is an application of the more simple method of consulting experts but provides a structure and a method of progressive refinement. Delphi, near Mount Parnassus in Greece, was the site of the most famous oracle of the ancient world. The Delphi technique was developed by the

Figure 3.2 **Technology forecasting methods**

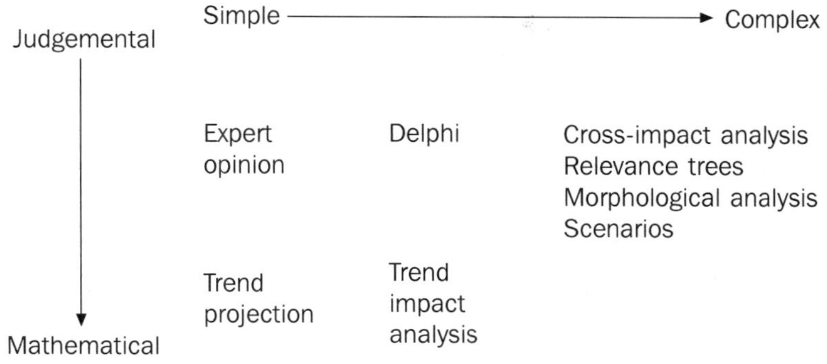

Rand Corporation to overcome some of the weaknesses of consulting with experts, especially in committees. It had been found that certain individuals tended to dominate such meetings and their views became the consensus view perhaps because others were unwilling to challenge them.

Committees also suffer from problems of finding convening times that suit all the members and of distance travelled and time out of the office required to attend. The Delphi method is to circulate a series of questions by mail to a panel of experts who are not necessarily aware of the identity of the other members.

In the first round, the co-ordinator sends questions to the panel members. A typical phrasing of a question might be 'Give description and a forecast of the market entry date of lightweight non-polluting energy storage systems suitable for widespread use in motor vehicles.' The respondents send their individual answers to the co-ordinator.

After analysis of the replies, the co-ordinator circulates the results and the respondents have a chance to reconsider their replies in the light of the opinions of the other panel members. Those outside the interquartile range are asked to give reasons for their replies, which might result from lack of knowledge or more importantly from specialist knowledge unknown to the other panel members.

In the next round, the results and analysis are recirculated together with reasons justifying the more extreme replies. The panellists are given a further opportunity to reconsider their views. The method encourages the experts to move towards a consensus, and further rounds can be carried out if necessary. The method can be difficult to organise but in-house Delphi exercises can also be useful provided there are enough experts. It can also be done by email. Sometimes a small number of outsiders are added to an internal panel. The technique can also be used to approach consensus on other subjects such as project selection and cost estimation.

Figure 3.3 gives part of a Delphi forecast for computer systems carried out by ICL in 1968. It is interesting to note whether the forecasts for the various events were correct or incorrect. Although some of the forecasts were correct the end result was achieved by PCs and not by mainframe computers as ICL had predicted. In fact they failed to see the impact of the PC and the Internet. This is another example of how difficult technology forecasting is and also shows how errors can made in both directions.

Information systems and sources

Modern information systems make it relatively easy to get information on a wide range of subjects at minimal expense. However, to search efficiently you need to know good sources of information and efficient ways to access them. The vast amount of information now easily available

Figure 3.3 Part of the Delphi forecast for computer systems (ICL)

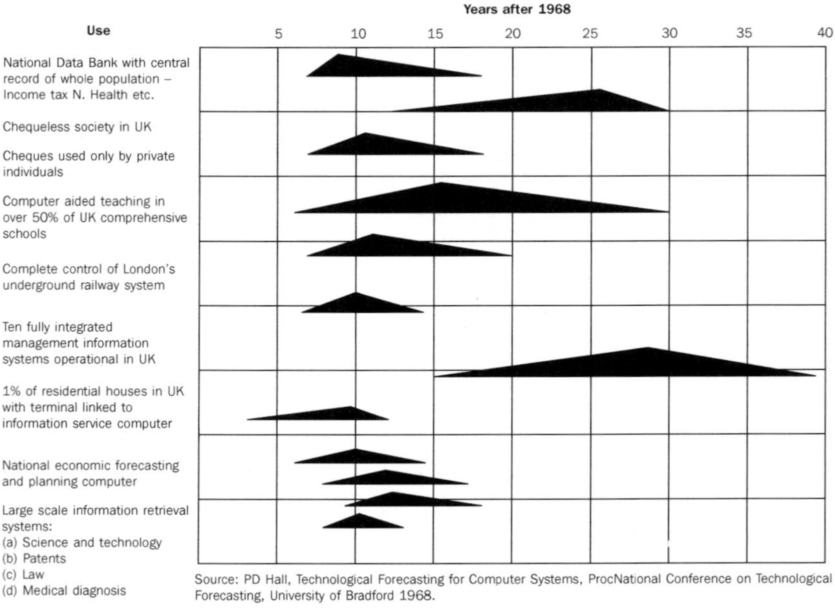

Source: PD Hall, Technological Forecasting for Computer Systems, ProcNational Conference on Technological Forecasting, University of Bradford 1968.

through the Internet often makes it harder to find important pieces of key knowledge without suitable training. This section emphasises the need for information searching and gives a short description of the general principles and development of on-line information systems.

Need for information searching

Many R&D projects are started on a hunch and large amounts of money are often spent on development before serious questions are asked about the project: Has it been done before? Is there useful technical information available on the subject? What is the 'state of the art' in terms of technology at this time? Is the technology available for licence?

Before a company embarks on any programme involving the introduction of new products or technology, it is essential to carry out an information search to find out what information is available on the subject and what work has been done before. For example, a patent search can show that a proposal to develop a particular new product may not be wise because it has already been developed and patented by somebody else. An EU report estimated that €20 billion has been wasted each year on repeating work that has already been carried out elsewhere.[4]

On-line information

In the past, it was often difficult to check out the market for new products or to check patent or technical information. Physical access to the appropriate literature was required. Now, however, most important literature is available from an on-line computer information service and anyone anywhere who has access to a computer and a telephone line can look up these information sources from their desk. Figure 3.4 shows how a computerised on-line information system operates. The user's PC is connected through a modem to an on-line service provider. This is a company that provides a data communication service for computer users which enables them to obtain information which is stored in computers in any part of the world.

Owners of large computers systems host vast databases of information provided by information providers. This content is usually licensed from publishers. Normally, users pay a subscription to the information providers if they need access to these specialist databases which usually contain commercially valuable information.

Figure 3.4 **Online information searching**

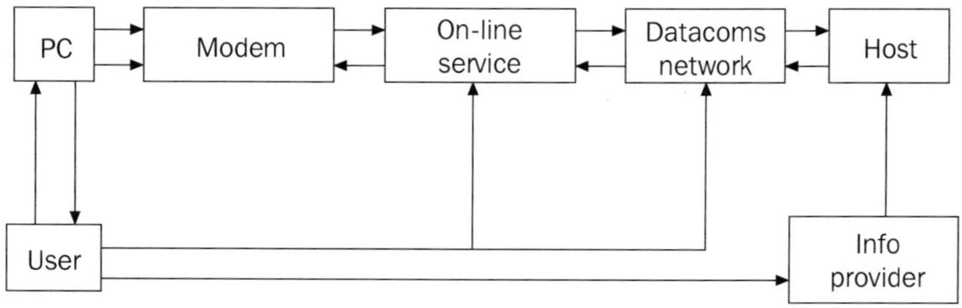

The availability of such services has evolved from the fact that computers started to be used for the editing and production of printed information sources such as technical journal abstracts. Once it was possible to store the printed information on computer, it was only a question of developing the necessary software so that this information could be searched in order to provide an information service. With the development of modems and computer communication it then became possible to access this information on-line from a computer terminal.

These subscription databases have been available since the 1970s. A more recent development is access to databases through the Internet. The subscription databases can be accessed through the Internet if the appropriate accounts are set up. It should be appreciated that subscription databases provide information that cannot normally be found by simple Internet searches. If you want detailed technical, patent or marketing information it is well worth subscribing to a suitable database provided you have a reasonable level of usage. However, there is a growing body of information on the Internet that is available free of charge, mainly due to public information policies of governments.

Using the World Wide Web you can research a topic very quickly. However, many of the search engines on the WWW are single level. You can put in a request and the computer will tell you that it can give you 23 000 references and will show you the first 10. What is needed is multilevel searching using logical operators to narrow down the number of hits to readable size. This type of facility is provided by specialist subscription databases listed below. Also, you can do some patent searching directly on the Internet. There are some free databases such as the European Patent Office (EPO), and the US Patent and Trademark Office (USPTO). These provide for the use of multiple search criteria to narrow down the number of hits to manageable proportions. The subject of patent searching will be covered in detail in Chapter 5.

Specialist databases

Examples are given below of a number of specialist subscription databases providing technical, market and patent information which are relevant if you are planning product/process development or searching for new product ideas. These databases are available from a number of information providers such as Dialog. You can gain some idea of the amount of online information available from the Dialog catalogue, which claims to have 15 terabytes of content from publishers all over the world.

A detailed list of useful web addresses for product and technology searching is provided in Appendix 1.

Table 3.2 Examples of subscription databases of interest in the introduction of new technology

Technical information	
Access to current research papers, technical literature and journals	
INSPEC	Electrical, Computing, Physics
ISMEC	Mechanical Engineering
CA SEARCH	Chemical Abstracts
FSTA	Food Science and Technology Abstracts
Market information	
Access to market data, trade statistics, company and financial data, business and economic journals and reports	
PROFOUND	Market Research
PROMT	Market Forecast Abstracts
FROST	Frost & Sullivan Market Reports
KOMPASS	Trade Directory – by Sector and Country
D&B	Dun & Bradstreet – Company Financial Data
Patent information	
DERWENT WPI	World Patent Index
INPADOC	Patent and Legal Status Information
DELPHION	(Formerly IBM patent server now Thomson)

Technology search techniques

A process of scanning of possible information sources is fundamental to any technology search. Some organisations employ a 'gatekeeper' to identify and circulate useful information. Other organisations carry out current awareness searches through relevant databases. These can also be done on a contract basis by technical consultants or universities.

The first step in carrying out a technology search is to identify arenas or hunting grounds. The search may begin with information obtained by technology forecasting, or the scanning techniques mentioned above. It can also result from brainstorming or other creative techniques used to introduce new and fresh approaches to company thinking. These creative techniques are described in Chapter 4.

Technology searching is also called technology scouting. This consists of carrying out a search for new products, processes or technologies. The term includes product and technology searching. Technology scouting can be done to meet the needs of a specific company or on a speculative basis to identify strategically useful technologies. Technology scouting is sometimes carried out by government agencies or on a multiclient basis for a group of firms.

A broad search could cover new technology areas. A narrower search could look for new product ideas within a certain area of technology. A basic and simple product search technique is introduced below. This can be extended to cover broader areas if required.

The product search philosophy

The basic philosophy in carrying out a product search is the same whether you intend to develop your own product or to look for a suitable licence or to purchase technology or a company. You have to initiate a systematic search through external sources of information with a view to arriving at a list of possible projects for consideration by the company. It is only after it has assessed such opportunities that a company can take a realistic decision about how to pursue the product opportunity: whether by development, licensing or copying as the case may be. The choice depends on the company's capabilities and also on proper assessment of the different opportunities so as to arrive at an appropriate strategy for exploiting each opportunity and market.

Research on innovation has shown the importance of marketing in new product introduction and companies cannot seriously expect to introduce new products unless the marketing function of the company is strong. Each company should assess its strengths and weaknesses in order to decide what features need to be strengthened, and to give pointers as to what areas the company is competent to tackle in terms of technologies and markets. Companies with established products and markets often have a good idea of what they want. They may have identified an opportunity and are looking for the product or technology that will help them to exploit it. Other firms may have difficulty in finding a starting point.

Finding a starting point

Finding a starting point can be a problem for smaller firms that do not devote enough resources to the strategic definition stages. Practical examples of starting points could be: the company's existing products, acquiring an agency to import a product and thus gaining market knowledge, entering into a partnership with a marketing firm or doing a survey of the company's customers to see what product requirements they have for the future.

Another approach is to look at growth trends in the market that may be caused by factors such as increased labour costs, environmental considerations, or developments and changes in safety regulations as mentioned in the previous chapter. It is also possible to identify market niches for special versions of products that your company can make and to fulfil particular requirements. Such products usually command higher prices.

Companies should also assess the experience that they have among their personnel, such as experience from previous employments and other markets. If you are still unclear where to start, you should start in an area you know something about and carry out the product search approach explained below on the basis that you can learn as you go along. The biggest mistake is not to start at all.

The product search process is an iterative process of progressive refinement of product or technology requirements. It is often hard to articulate your exact needs, but most managers are able to express a view on specific product or technology opportunities when these are presented to them. The process works by building up a profile of interesting product and technology areas by successive refinement of the company's needs as it goes through a repetitive search that zeroes in on suitable prospects. A summary of the process is as follows:

- Identify search arena.
- Identify suitable sources.
- Get information from sources.
- Screen, re-evaluate search criteria.
- Repeat search.

Technology sources

There are a number of approaches in sourcing technology or product ideas. These are based on geographical considerations or on the method of communication. You might think that it is better to start looking close to home, but in fact what you are looking for is something that does not have a strong presence in your market, so it is better to look as far afield as possible.

Technology sources by country

Cultural and language differences constitute an important barrier in technology searching. It is not surprising therefore that North America is a very productive area in which to search for new technology. An analysis of the membership of the Licensing Executives Society shows that 40 per cent of its 9000 members are in the USA. The US runs a huge surplus on technology exports over imports. In 1996 the US received $32.8bn in licence fees while paying out only $7.8bn.

Far East countries are significant developers of technology, but cultural and language barriers make technology transfer deals more difficult to achieve for English-speaking searchers. The EU itself is another source worth looking at, and a number of successful deals have been concluded between EU firms with assistance from the EU funded Innovation Relay Centres.

In general, technology transfer from companies within the same geographical region is only successful if the product or service is limited geographically. There is good potential for the licensing of construction products that are expensive to transport or software products that have specific application to one country.

Sources by contact method

The many different sources of new products fall into two basic categories: direct contacts and literature or documentary sources. Direct contacts involve talking or writing to people and therefore involve some degree of disclosure of the company's interest or intentions. By contrast, you can consult literature or documentary sources without revealing the name of your company and this has its advantages, especially in the evaluation stage of a project.

Direct contacts

Direct contacts involve various sources, some of which can be contacted both at home and abroad. In addition, there are a number of sources

Table 3.3 External product sources

Direct contacts

At home:

- Research institutes
- Industrial development bodies
- Innovation centres
- Consultants
- Inventors
- Universities
- Embassies
- Customers

Abroad:

- All of the above plus
- Foreign firms
- Trade fairs

Literature or documentary sources

- Trade directories
- Industry associations
- Industry/trade magazines
- Patent literature
- Venture capital sources
- Licensing networks
- Company catalogues

abroad of which the most important one is foreign firms. Taking, for example, a country such as the USA with a population of 293 million people, it is not generally appreciated that only 10 per cent of US manufacturing companies export, and that there is a huge untapped pool of small- and medium-sized US companies that are not represented in Europe in any way. By identifying and approaching these companies directly, it is often possible to arrange a licence simply on the basis that the company had never been approached by anybody before and is flattered at the thought that somebody would pay them money for the use of their technology. Many of them hold patents but these patents have no European equivalents. Smaller firms relate well to similar-size firms in the UK and Ireland and may be delighted to be approached since they have no prospect of direct entry to the EU market. Development work may be required to convert products to European standards. Possible firms are usually identified through the published sources or through trade fairs.

Trade fairs

Trade fairs can be very useful for identifying certain types of firms. They are particularly suitable for identifying established firms with new product offerings. At a trade fair, you can see which stands have unique products and which have products that are run-of-the-mill. Fairs are a good source for engineering products in particular. As there are many specialised trade fairs in the USA, it is possible to be quite specific in your search.

Trade fairs and exhibitions are also of considerable value in finding out what is happening in an industry. However, you should not expect to be able to enter into licence discussions on an exhibition stand. Obtain an introduction and make an appointment to see the correct executive at a later time. Technology fairs are of limited value because most exciting technology is licensed directly without appearing at such fairs. However, fairs are a useful source of making contacts for future reference.

Consultants and brokerage

There are only a few successful consultants in the technology transfer area in Europe. Some consultancy activity that was funded by the European Commission has now ceased. Technology brokerage on a no-foal no-fee basis has not yet become a viable business in Europe and most consultants work on a fee basis or a mixture of fees and commission or royalty. Usually they are commissioned by a potential licensee to search for technology, and no fee is paid by the licensor. Brokerage on a commission basis is more common in the USA, where there is a bigger market. The recent development of Internet-based brokerage is making the activity more profitable.

Foreign trade missions or embassies

Many countries maintain technology scouting and brokerage activities through their foreign trade missions in other countries. They also provide trade links to firms in their own countries, usually to encourage exports but also to export technology. It is usually best to approach them on the basis of looking for products to import rather than asking for licences on first contact.

Research funding organisations

Traditionally, research funding organisations such as British Technology Group have funded the development of university and other inventions submitted to them in return for ownership of the intellectual property rights and a royalty split with the inventors or university. The model has changed especially in the USA to one of identifying strategic technologies, finding suitable researchers and funding the projects. The projects are then commercialised through equity funded start-up companies or licensed to target prospects. This means that research organisations do not have the same level of technology freely available for licence, and access to technology of this kind is better sourced direct from the universities or through venture capital sources.

Universities

These are a rich source, especially in the USA. Their technology is usually at an early stage of development and only suitable for companies that can complete the development and assess the technical and commercial risks. A useful source is the Association of University Technology Managers, www.autm.net.

Research institutes

There are 700 government research institutes in the USA with funding of over $60bn. Their lack of success in transferring their technology to industry in the USA has been a cause of concern for the US government and many initiatives have been put in place to assist the process. They tend to patent their developments only in the USA and publish lists of patents for licence. In general they do not expect to be approached for licences from outside the USA and defence considerations can in some cases limit their room to manoeuvre even where the defence applications are very limited. Research organisations in other countries are more open in their approach to international licensing. Like the universities, their technology is often at an early stage of development.

Literature or documentary sources

Published sources are indirect sources and are a means to identify the direct sources such as companies and research organisations such as universities. The Internet is usually the starting point for published sources and Internet searches are becoming more productive especially where they lead to other information resources such as trade directories, databases of universities or industry associations. (See Appendix 1.)

Trade directories

These are ideal if you have a clearly defined requirement or industry sector that can be identified through the categories indexed. They are more suitable for a product than for a technology search so the product-searching route should be pursued as the way to identify suitable technology. Thomas Register and Kompass are useful and are available on-line.

Industry associations

The USA has many industry associations that provide useful leads to companies in an industry and can give addresses and websites. These can be identified these by reading the US trade magazines or through the Internet.

Trade magazines

Trade magazines are essential reading for the technical commercial specialist. They include new product announcements, debate issues of concern to each sector and have reply-paid cards that you can use to contact companies and collect intelligence on the products and prices without revealing any predatory intent.

Patents

Patents are the biggest source of technical information available, with over 1 million patent applications being filed each year. Patent applications include much technical know-how, which is provided as background information and does not form part of the patent claims. Many inventions have very limited patent coverage, for example many US small firms do not file applications in Europe and so these inventions can be copied provided the product is not sold in the USA. Much US government research is communicated to industry by filing patents and very little of this research is protected in Europe.

However, looking at patents does not inform the searcher if a licence is available. It is necessary to contact the patentee to find this out. If the company operates globally there is much less chance that rights to Europe will be available. Database information now lists patent families, so it is easy to see if patents have limited territorial coverage. It is also quite easy to search by date so as to find patents that are soon to expire.

Patents often provide the earliest publication details of a new invention or technology and with most patents being published within 18 months of first filing or earlier in some cases. These publications usually pre-date any trade or technical publications by some time.

Venture capital sources

In the USA, many companies reveal their technological intentions and developments when looking for venture capital. Some of the venture capital databases, such as www.venturewire.com and www.venturesource.com (the latter being subscription only) are an early source of information on new technology. These give addresses and websites of companies worldwide that have secured funding in the USA.

Licensing Executives Society (LES)

The main advantage of LES in terms of searching for technology is the access its members get to its Members Directory. This contains the contact details of the licensing managers of most large US companies and almost all of the US universities. This information occupies about 40 per cent of the directory. The other 60 per cent is devoted to coverage of the rest of the world on a less comprehensive basis. Smaller companies are not well represented in the LES, and many of the members are company lawyers.

Licensing websites and networks

The Internet is the ideal mechanism for connecting licensors and licensees. The Innovation Relay Centre network is a good example of how such a system can work. This network funded by the EU gives access to a large amount of EU-funded projects. Internationally, the licensing website, yet2.com is delivering good results at www.yet2.com. It also provides facilities for browsing if the search is broad. Another interesting website is http://pharmalicensing.com

Carrying out the search

The stages in carrying out the product search are shown in Figure 3.5.

Having reviewed the sources of product or technology opportunities, it is important to assign priorities and staff to the various sources in order to make sure that scarce resources are correctly allocated and that people do not spend time looking at areas that may be of less value to the company. It is important then to decide on how contacts will be made. This involves deciding on how the company will present itself, and writing a product search brief. It is wise to have a folder that describes the company and provides some corporate image information. Into this folder you can place sales leaflets giving details of various products that may change from time to time. The product search brief consists of a page that is also inserted into the folder that describes the company's broad objectives in relation to new products and links them to various strategic objectives such as that of growing within the Single European Market. It is important not to be too specific in stating product

Figure 3.5 **Stages in the product search and licensing process**

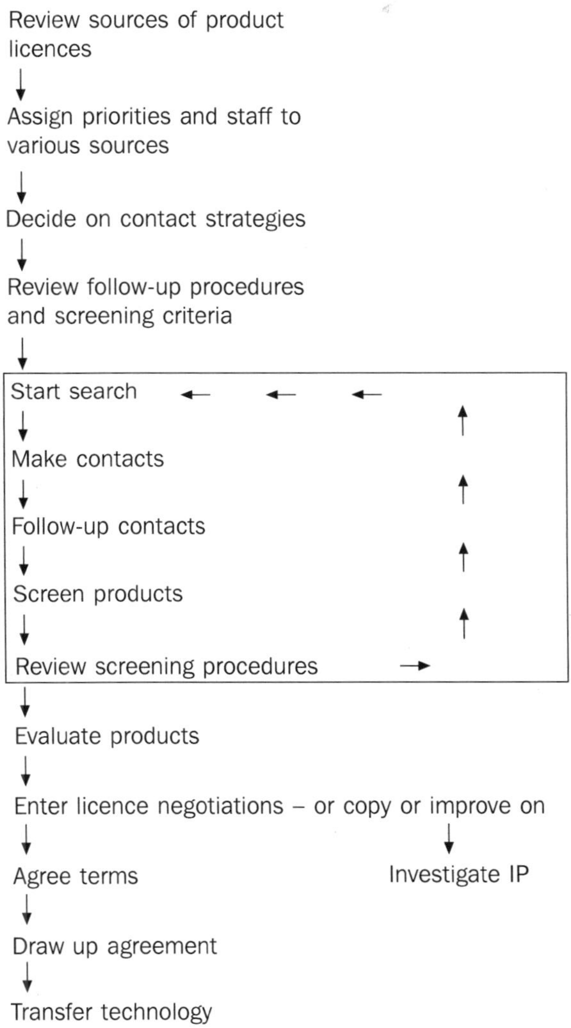

requirements, as the company that you contact may not have exactly what you require but could have something much more attractive that you didn't know about.

One person should make all the contacts and should be in charge of reviewing the information received. It is also of value at this stage to decide on screening criteria to filter the incoming information. This is usually done by setting 'Must Meet' criteria for projects. Such criteria are also used to evaluate initial proposals for R&D projects and a more detailed treatment of this subject is given in Chapter 5. In summary, a list of criteria that any project must meet in order to be considered further is drawn up. Failure to meet any one criterion rules a project out.

Once the search begins and contacts are made, information is received from the follow-up which must then be screened according to the must meet criteria. Usually you find that the screen has either been set too coarse or too fine. If the screen is set too coarse, it means that the company is considering too many possible projects and it ends up with an impossible evaluation load. If the screen is set too fine, the company has set very narrow criteria and very few suitable opportunities will be found. By going around the loop, you get a better idea of what is on offer and the screening criteria are constantly re-evaluated in the light of experience.

In a product search, very few people know exactly what they want at the beginning, but when presented with information on possible opportunities they are able to express an opinion as to whether the opportunity is of interest or not. In other words, it is possible to build up a template of acceptable projects by a process of successive refinement. The product search process works on the basis of presenting a number of opportunities to the company in a structured way and refining the screening criteria in a formal way so that the searching process is constantly being refocused and the company's criteria for areas of interest are gradually being more accurately defined.

Contacting large or small companies

Experience in licensing has shown that companies should try to do business with other companies of the same size. In looking for a licensee, a large company will automatically expect to find a licensee that is of similar size to itself. It will be unwilling to entrust its technology or products to a company that, by its standards, is much smaller, financially weaker and not very well spread geographically. A small company looking for a licence will find that its problems will be better understood by other small or possibly medium-sized companies abroad. They will not have such high expectations in a licensee, and in many cases if a company is being approached for a licence for the first time, they may not have any expectations at all. It is possible for the potential licensee to build a relationship of trust on a personal basis with the owner/manager of a small company and as licensing involves long-term relationships, the personal aspects are very important.

Smaller companies can be more open in carrying out such a search as their geographical spread is small and they are less likely to be contacting a potential competitor. Larger and multinational companies will tend to collect larger amounts of information without revealing who they are and will carry out some evaluation before making a direct approach.

Evaluation and review

The search exercise is usually carried out over an extended period with review meetings to screen the incoming opportunities and decide which should be subjected to detailed evaluation. The subject of evaluation is covered in Chapters 5 and 12 but will also require a detailed knowledge of intellectual property, which is covered in Chapters 7–9.

Depending on the resources available, a search should go on for a period of a few months as a part-time activity with the objective of presenting a number of possible projects for screening at review meetings held every 2–4 weeks. As the search progresses, the individuals involved reach a point where they become familiar with all relevant technologies and players in the area examined. This happens after the search loop has been traversed many times. The method is an iterative process of zeroing in on possible opportunities. It should not be restricted to licences and should cover other opportunities that may or may not be available for licence. Part of the evaluation, as we shall see later, is to decide how to acquire the technology.

It is important to remember that the product or technology search exercise is useful before deciding on the exact technology acquisition strategy. A company may intend to enter an R&D programme, but carrying out the search could cause them to find that the technology they need could be licensed or even copied if it is not protected. The outcome of the evaluation stage will be a go/kill decision to acquire or not to acquire the technology. If the decision is to acquire the technology, the next decision will be to decide whether to acquire it by internal development or to source it externally.

In some cases a combination of internal and external sourcing is the best solution. This is particularly common in the development of software that may use specialist components. For example a new computer game may be developed and a component such as a physics engine incorporating the laws of motion may be licensed-in and incorporated into the product.

Google Earth: an example of technology acquisition

Google conducts regular searches for technology and in 2004 it announced the acquisition of Keyhole Inc. a start-up company founded in 2001 that sells software for viewing 3-D maps and images from as far away as space or from close to street level. The privately owned company received its initial backing from Sony Broadband with further funding from In-Q-Tel, an investment arm of the Central Intelligence Agency. Jonathan Rosenberg, vice president of product marketing at Google, said in a statement: 'Keyhole is a valuable addition to Google's efforts to organize the world's information and make it universally accessible and useful.'

Google has now launched Keyhole's software product in the form of Google Earth a free downloadable package that can be used to view any point in the world from above so as to find a local hospital, hotel, restaurant or business in locations around the world.

Notes

1. Cooper, R.G. (2001) *Winning at New Products*. Perseus, pp. 11–12.
2. Prehoda, R.W. (1967) *Designing the Future – The Role of Technological Forecasting*. Chilton.
3. Twiss, B.C. (1980) *Managing Technological Innovation*. Longman, pp. 206–234.
4. *Packaging Week* (1997) Jan 23: 17.

4

Idea generation

Introduction

Ideas for new products and technologies to be developed by a company can result from external product or technology searches as we saw in Chapter 3, or they can result from internally generated sources. In many cases, the technology acquisition strategy may not be finalised until the external product or technology search exercise confirms that the technology required cannot be obtained externally. A decision is then taken to embark on a development programme – the acquisition strategy of inventing something completely new or improving on previous technology. The advantage of doing the external search first is that the development programme when carried out will start from a much more informed perspective.

This chapter identifies all the internal sources available to the company, describes how to stimulate the sources, and how to manage the idea generation process.

Idea generation activities

Gaining the maximum from internal idea sources breaks down into two main activities:

- Identifying and tapping sources within the organisation.
- Stimulating the creative processes and behaviour.

Internal sources

It is good practice to assign one person to the job of identifying and tapping internal sources. This person should be approachable and open to new ideas, and might also function also as the gatekeeper identified earlier in Chapter 3. The main internal source is the employee suggestion scheme. By setting up such a scheme the company gives out the message that it is looking for ideas and is prepared to reward and encourage people to be helpful and creative. This involves setting up a mechanism for receiving and evaluating the ideas submitted.

To operate a suggestion scheme, the company must set up a permanent evaluation committee, which must meet regularly to assess the ideas and decide on the next steps to be taken. This can be the same committee which applies 'must meet' criteria to project submissions. Suggestion schemes work best if they are seen to be open to all staff, not just technical or marketing staff, and open to all kinds of suggestions from new product ideas to reducing the queues in the canteen. In a more technical environment it may be possible to restrict suggestions to new product or process ideas. In encouraging suggestions there is no harm in advising staff that the acceptance rate of ideas suggested may be low. Ideas submitted should be acknowledged immediately by the secretary of the committee and the proposer informed of the date of the next review meeting.

The evaluation committee should decide on the ideas quickly and inform the person who submitted the idea promptly and tactfully of its decision. It should not wait until a decision has been taken to proceed with the idea as this can take some time. The proper response is to communicate the outcome of the first review. This will be to reject the idea or to carry out a detailed assessment. Thus the submitter is told that the that the proposal was rejected or that it will be assessed in detail before a final decision is taken. Implementation is often sufficient reward, but cash rewards are also common. Making rewards too large or giving excessive praise can result in jealousy.

As well as running a suggestion scheme that is open to all employees, it is recognised that certain people act in a more creative or information gathering role than others and mechanisms need to be put in place to make sure that the ideas of these people are tapped. Thus the new product introduction process should include regular consultation with sales and marketing staff and with customers.

As well as talking to the right people, new product ideas also come from looking in the right places. For example, carrying out a detailed and critical examination of the company's existing products and comparing them with competitors' products, or looking for gaps in the product line can produce fruitful ideas.

There are a number of techniques that can be used to analyse and solve problems and to stimulate the creative processes. These include creating a suitable environment for creative problem solving and encouraging creative behaviour. It has been shown that people are more creative if they are able to work in a supportive environment.

Quantity over quality

The importance of generating large numbers of ideas is indicated by the low success rates for new products discussed in Chapter 3. If only one in seven product concepts and one in four development projects are successful we need many ideas at the input stage if we are to ensure success. There have been a number of studies that have shown that a large number of ideas are needed to ensure a successful outcome. Greg Stevens[1] has suggested that 3000 raw ideas are needed to ensure one commercial success. We have to be careful how we define ideas for new products and raw ideas may need considerable filtering before they become product concepts. An earlier study by Booz et al. in 1982 suggested that 58 ideas were needed to get one successful new product out the end of the process.[2] But as businesses have come to focus their idea generation more closely towards their strategic objectives the number of ideas needed has fallen such that Cooper reports a one in seven success rate for product concepts.[3]

Management are often reluctant to devote resources to idea generation, regarding it as inefficient and an excuse for time-wasting. However, research clearly shows that certain creative employees deserve to be given the freedom to come up with lots of ideas, and that the process should only be judged by the end result after some time and not by the fact that crazy ideas may be produced along the way. A crazy idea may be the stepping stone to a commercial winner.

Problem-solving techniques

The selection of techniques for solving problems is based on the nature of the problem and how it is stated. Some problems have only one correct solution, for example to find the cost of a new product design. The solutions to such problems use analytical techniques that involve a series of logical steps, each of which has only one correct answer. Such problems are described as closed, and the thought process involved is described as sequential or vertical.

Other problems may have more than one correct answer. For example, how to reduce the cost of a new product, or how to come up with new designs. These are called open-ended problems. These kinds of problems can be solved by using techniques that are less analytical and more creative. They cannot be solved by techniques that try to zero in on one correct answer. The thought processes are not necessarily sequential. One idea is not necessarily built upon another.

Some open-ended problems may be stated in a manner that restricts the possible solutions, and so techniques may be employed to restructure or redefine the problem. In a design group someone might ask, 'Why can't we reduce the weight of this component?' This approach closes the problem, so a better approach might be to ask, 'What can we do to improve the size, shape, weight or function of this component?'

Creativity

Creativity is the ability to relate things and ideas that were previously unrelated. It implies breaking the more logical thought process and introducing a more intuitive mode of thinking.

Creative thinking

Conventional thinking is logical, sequential or vertical and is associated with the left side of the brain. Creative thinking is illogical, parallel or lateral, and is associated with the right side of the brain.

Creativity is described by Edward de Bono[4] as a result, while creative thinking is described as a process. A result can only be admired, he says, but you can learn to use a process. 'Lateral Thinking' is the name de Bono has coined for his approach to creative thinking. He describes the process of lateral thinking as one that should not be confined to the idea of inventing things but is a method of generating new ideas that can be applied in every field, from science to business and even in one's personal life.

Lateral thinking can be used to enhance the process of vertical thinking. New ideas produced by lateral thinking can be developed further by vertical thinking. The traditional education process tends to emphasise vertical thinking. As a result, people are conditioned to think vertically, a process that excludes possibilities at every stage, constantly rejecting ideas that could have merit. One of the key steps in creative thinking is separating idea generation from evaluation.

The effect of traditional education methods can be seen in the way children ask questions.

Figure 4.1 The three ages of man (and woman)

- 0–5 years – Why?
- 5–10 years – Why not?
- 10 years+ – Because

Revert to childhood (temporarily)

Our normal vertical mode of thinking is based on 'Because'. We are constantly narrowing down possibilities and looking for early answers. We say that it won't work 'because'. Premature judgement is one of the biggest killers of creativity. The suggestion is that we should revert to childhood, go back to the 'why not' stage and be less willing to settle for the safe, easy and less creative solution.

Suspended judgement

De Bono emphasises the importance of suspending judgement. The purpose of creative thinking, he says, is not to be right but to be effective. 'Being right means being right all the time. Being effective means being right only at the end.'

To avoid judgement, de Bono suggests the use of a new word. If we are not sure about something we normally say 'NO'. De Bono suggests that we say 'PO' instead. In other words we have not yet allowed any rejection of the idea to take place. We are not saying yes, but we are also not saying no either. PO allows us to break out of 'digital thinking' where there are only two possibilities, as in a computer.

A group of children was asked to come up with a design of machine for going over rough ground. One child suggested a machine that had a reservoir of 'smooth stuff' at the front, which it spread out in front of it, ran over, and then collected again at the back. It would be easy to dismiss this concept as naïve but before you do that you should ask if there is some way that it could be made to work. Maybe then you might come up with the idea of caterpillar tracks. It is important to recognise the usefulness of 'the intermediate impossible', an idea that does not work but which should not be dismissed, at least not yet. Remember, the objective is to be right but only at the end.

Another group of children was asked to design a dog exercising machine. One child came up with the design shown in Figure 4.2.

We should ask ourselves what was the creative way of looking at the problem that enabled the child to invent this machine? The answer is that instead of thinking of ways to make the dog go, this inventor thought of a way to stop the dog from stopping. When the dog stops, the momentum of the trailer causes it to catch up with him and he gets an electric shock. This is an example of how restating a problem can provide a creative solution.

Figure 4.2 Dog exercising machine

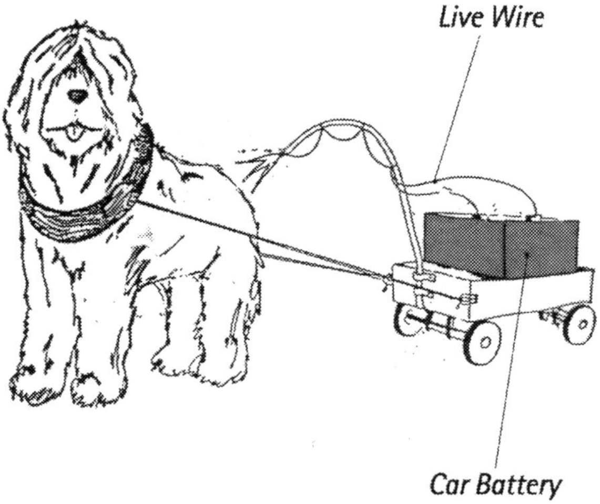

Creative behaviour

Teresa Amabile[5] states that managers can influence three components of creativity: expertise, creative thinking skills and motivation. She reports that her research found that people, appreciate challenges where they are matched to the right assignment, have the freedom to decide how best to do the job, and are given adequate resources to carry it out.

Creativity is also encouraged by supportive management. People are motivated by encouragement and by limited use of rank and authority. In creative sessions staff should not feel inhibited by the presence of senior personnel. Encouragement comes from reinforcement of ideas, from seeing one's ideas being supported or by others submitting similar ideas.

Barriers to creativity result from inhibition, people being afraid of making a fool of themselves, being afraid that they or their suggestions are being judged, or submissiveness in the presence of superiors. People should not be punished unless their performance is poor over some considerable time. In development work it is important to allow people to be wrong so that they can be right at the end.

Applications of creative thinking

Creative thinking can be applied to all areas of life. This book concentrates on the application of creative thinking to the acquisition of new technology. In particular, creative thinking has specific applications in product development. These range from generating new ideas to solving technical problems.

Idea generation

As we have seen earlier, new ideas can come from market pull or technology push, and business and technology strategies try to integrate these idea drivers to ensure that the arenas or hunting grounds chosen line up with the long-term development plans. The purpose of idea generation and the use of creative techniques is to arrive at a product specification or market opportunity.

Figure 4.3 Idea sources in product development

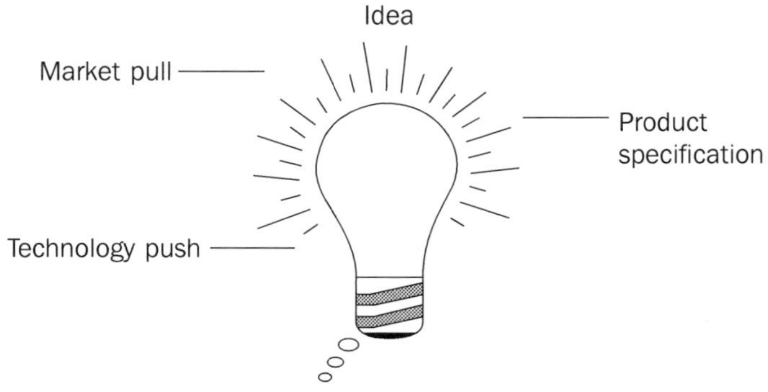

A product specification is a definition of a product or market opportunity that specifies what the product should do without specifying how it should work. There may be a basic technical principle in mind, but the best method of idea generation is to keep the idea stage free from any constraints or technical bias so as to allow for maximum creativity.

A company that manufactures springs might set up an idea generation exercise stated as follows: 'Generate new ideas for new products which will make use of springs.' While this is a good use of creative techniques, it is based entirely on technology push and a more integrated approach that does not lock in a technological solution is better. It could be that a more in-depth business analysis might show that the manufacture of springs using current technology had become too competitive and that the company should seek to broaden its technology base and develop a more market-driven approach to product development.

Solving technical problems

The separation of idea generation from any technical constraints allows for a second use of creative techniques in turning raw product specifications into product concepts. Product concepts are ideas that have had a possible technical solution added.

Figure 4.4 shows how creative techniques can be employed in the stages of product development to go from product specifications through to product concepts that are then evaluated and developed. The figure also shows other possible useful inputs to concept generation coming

Figure 4.4 **Creative inputs to product concepts**

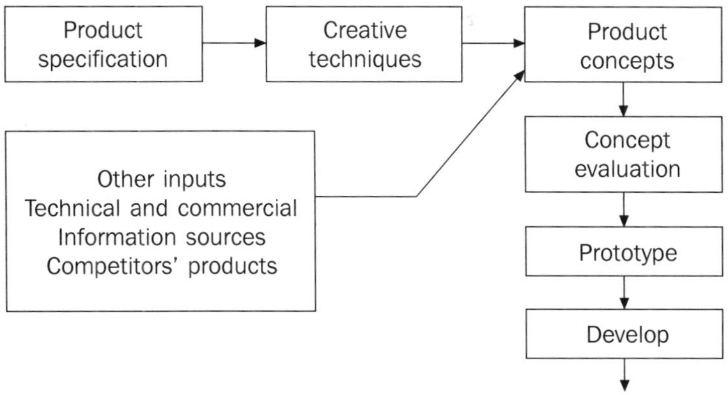

from competitive analysis and technical and commercial information sources. In addition, there are opportunities to employ creative techniques throughout the technical development stages and right through to market launch.

Stages in creative problem solving

Problems can be solved in a more creative way by going through a series of stages. The stages apply a disciplined approach to problem solving.

1. *Preparation*. Obtain sufficient information about the problem.
2. *Definition*. Define the problem in a number of different ways.
3. *Looking*. Apply techniques to find solutions.
4. *Incubation*. Take a break: allow your subconscious to work on the problem.
5. *Repeat*. Have another look at the problem.
6. *Evaluate*. Ask: are the ideas any good?

Individual techniques

Some techniques work well on an individual basis. The same techniques can be used in groups but there are group techniques that are very powerful as they take advantage of the creative atmosphere that can be generated in a group and this leads to cross-fertilisation. People build on each other's ideas and the net result is much greater than what people could achieve on their own.

Although group techniques work well, there are many occasions when individual techniques are more appropriate. Often it is not practical to call a group together, or a solution is needed quickly or there is only one problem to be solved and the owner of the problem is much more familiar with it than anybody else. It helps to avoid groupthink – the tendency to agree with the dominant figure in a group.

Problem redefinition

This technique was illustrated in the designing of the dog exercising machine above. You must step back from the problem and ask if it can be stated in another way. First look at all the words used and make sure you understand them. Ask if the problem has been stated in its simplest form. Take any of the key words and ask 'What does this mean?' Try to remove any associations in the use of the words to describe the problem.

A group is asked to suggest new methods of preventing car theft in a city. The problem can be redefined by stating that it is not so much that people's cars are stolen but that they are deprived of the use of them. They would have the same problem if the car broke down or was towed away by the traffic police. While the immediate solutions to the problem as originally stated would be to develop better security devices, the restatement of the problem suggests solutions such as persuading people to leave their cars at home by providing better public transport, or encouraging car-pooling by rewarding car pools with low cost secure off-street parking.

Osborne's checklist

Osborne's checklist is used for changing the attributes of the problem or object being studied:

- Other uses?
- Adapt?
- Modify?
- Magnify?
- Minify?
- Substitute?
- Rearrange?
- Reverse?
- Combine?

The purpose of the checklist is to encourage other ways of looking at the problem. It forces new ways of thinking about an idea. You are asked to consider how the problem might be solved or the idea enhanced by making it bigger or smaller, for example. The use of the checklist can become automatic after some time.

Inversion

This technique asks you to think about something and then turn it inside out or upside down. You can also turn it backwards or sideways. The technique is best illustrated by the example shown in Figure 4.5. This shows a design of a pneumatic cylinder that can be made up to 200 metres long!

Called the Peristaltic Actuator this product was invented by Bob Rees. He was asked to develop a pneumatic cylinder for opening the doors of an aircraft hangar for jumbo jets. A Boeing 747 has a wingspan of over 60 metres, so a long cylinder was needed. Conventional pneumatic cylinders are seldom made in lengths over 2 metres. They have a piston rod so when they are extended they become twice as long as when they are retracted. In their retracted form they are half the length of their operating stroke. Even if you split the hangar door into a left and right, you would still have a piston rod sticking out 40 metres from the side of the building.

Using the inversion technique, Rees came up with the idea of turning the cylinder inside out, of putting the piston on the outside. The cylinder was made from fire hose, which is very durable, folds flat and can be up to 200 metres long. The piston was made in the form of a set of two pinching rollers. Air is pumped in at one end of the hose and the rollers move as the hose tries to become circular. The leading edge of the door

Figure 4.5 Peristaltic actuator

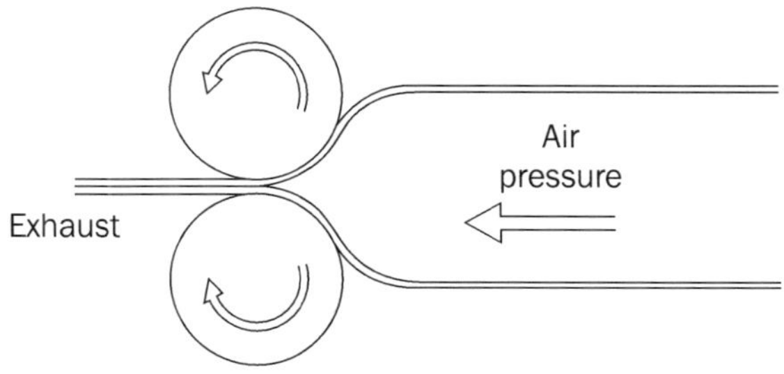

is attached to the outside casing of the roller carriage. The product is now successfully used in many long stroke cylinder applications.

Ideas and words

This technique is useful if you are stuck on a problem and want to look at it in a new way. Write down the key words in the problem and then use a dictionary or thesaurus to find other words of similar meaning. One of these synonyms might help to spark off new ideas and associations that help to solve the problem. You can do this on-line at www.thesaurus.com.

Morphological analysis

This technique forces new combinations of elements in the form of a device under consideration. The name is based on the Greek word 'morphe', which means form. A clock has a certain form. We may consider it to have a method of time indication, say a dial, a method of mounting the dial, a mechanism for time regulation and a power source.

An ordinary clock on the mantelpiece might have a circular analogue dial mounted on a wooden casing with a spring escapement to regulate the time and be driven by a spring. This is one form. We can take the elements of the clock and try to list alternatives, as shown in Table 4.1.

The more obvious forms are shown in the upper part of the table. More unusual forms are shown lower down. To carry out this technique, the items in each cell of the table are written on cards and arranged in four piles, one for each of the elements. The piles are then shuffled individually and the cards from each pile are laid out vertically to create a new table. It can be seen that this will create new combinations of elements. In this case the number of combinations is $9 \times 8 \times 9 \times 10 = 6480$ different combinations.

Mind mapping

This technique was invented by Tony Buzan as a method to help learning. He found that people learn things more effectively by taking notes where they write down key words. He also found that people have

Table 4.1 Morphological analysis of a clock

	Elements of clock			
Forms	Time indication	Mounting	Regulation	Power sources
1	Analogue dial	Simple casing	Spring escapement	Clockwork spring
2	Chimes	Floor standing	Pendulum escapement	Weights
3	LCD digital display	Wrist	Crystal	Battery
4	Digital number plates	Shelf mounted	Mains frequency	Mains electric power
5	Voice synthesiser	Wall mounted	Atomic signal	Solar battery
6	Vibration	Pocket	Momentum	Walking movement
7	Electronic tubes	Instrument case	Oscillator	Battery/electric
8	Projection on wall	Outdoor case	Isotope decay	Wind power electric
9	Linear display		Tuning fork	Water powered
10				Temperature change

excellent visual memories and by writing diagrams of key words, people not only could retain the information but the technique could also be used to encourage creative thinking.

Mind maps are diagrams that are drawn around a central idea or key word. If you ask two different people to draw a mind map around the same key word, their answers will be very different giving rise to many new ideas. Figure 4.6 shows a mind map drawn around the idea of finding ways to promote a newly published book.

A number of ideas have been drawn as lines coming out from the central idea. One creative idea shown was to use spare covers that the printer had delivered. Another idea was to mail leaflets to bookshops. This idea was further explored and it was decided to use the spare covers as a promotional leaflet.

Figure 4.6 **Mind map for promoting a book**

Group techniques

Group techniques can be much more powerful than individual techniques because they provide an opportunity for cross-fertilisation. One person can put forward an idea that can be built on by the other people in the group. Another advantage of groups is that an atmosphere or psychological state is created where people let go of their inhibitions and ideas flow freely. The main features of group creativity sessions are that a group of suitable people is assembled in comfortable surroundings free from interruptions, where they are encouraged to forget about external problems and just focus on the problems presented for creative solutions by the group leader.

Brainstorming

The most popular group creative technique is brainstorming. Unfortunately the term is often misused and has come to be associated with any type of group problem-solving session. To work correctly, it is essential to keep to the basic principles of a formal brainstorming session. These are shown in Table 4.2.

A typical brainstorming session should run for about 20 minutes per problem. After this, ideas begin to dry up and the atmosphere evaporates. One of the factors that can inhibit the session is the length of time it

Table 4.2 **Elements of a formal brainstorming session**

Appoint facilitator	
Define problems to be addressed	Stated in simple form Free of associations
Select group	Between 5 and 10 people Of similar rank and status With a mix of skills and backgrounds
Create atmosphere	Comfortable surroundings Free from interruptions Preferably off site Encourage humour
Display ideas	Flip chart or Cards
Postpone judgement	Cut off judgemental comments

takes for the facilitator to write the ideas on the flip chart. If the ideas are coming thick and fast the momentum can be lost if the facilitator has to keep asking people to repeat their ideas. One way to avoid this is to give out cards to the participants and to ask them to write their ideas on the cards. These can then be stuck to the flip chart or laid out at the centre of a table where everyone can see them.

The role of the facilitator is to encourage the participants and record their contributions. Sometimes the facilitator needs to suggest a few ideas, remind people of some previous ideas and ask them to think about them or move people on if they have got stuck in a narrow rut. The facilitator should be ruthless in stopping any judgemental comments. The key to successful brainstorming is to separate the idea generation and evaluation activities. All ideas should be written up no matter how crazy. A crazy idea may be the impossible intermediate that sparks off another idea that turns out to be a winner. Brainstorming is particularly suitable for the generation of simple ideas such as suggestions for new products. It is less good at generating solutions to technical problems.

Synectics

Synectics is a technique developed by William Gordon. The term comes from the Greek word 'Synectikos', which means bringing forth together. Gordon learned from his researches into creative thinking that creativity is associated with certain psychological states. Synectics was developed as a systematic way to encourage these states in groups. This is brought about by the use of metaphor and analogy in which people are encouraged to engage in flights of fancy.

Initially the technique can be used informally to develop active listening and encourage contributions by paraphrasing ideas, giving feedback and making constructive evaluations. At a more formal level, trigger questions are introduced to encourage new approaches, or the group is led on an excursion. This is a process where some unrelated idea is introduced to the group, which is then asked to engage in speculative thinking where the new idea might create completely new insights.

Notes

1. Stevens, G.A. and Burley, J.R. (1997) 3,000 raw ideas = 1 commercial success. *Research Technology Management* 40(3): 16–27.
2. Booz, Allen & Hamilton (1982) *New Product Management for the 1980s*. New York: Booz, Allen & Hamilton Inc.
3. Cooper, R.G. (2001) *Winning at New Products*. Perseus, pp. 11–12.
4. De Bono, E. (1990) *Lateral Thinking: A Textbook of Creativity*. Penguin.
5. Amabile, T.M. (1998) How to kill creativity. *Harvard Business Review* 76 (5): 76.

5

Evaluation of new technologies and projects

Introduction

Having dealt with the ways of acquiring technology, we can now go on to consider how to decide on the selection of the particular technology or project that will best fit it with the company's business and technology strategy on an ongoing basis. The principles described in this chapter relate to technology in general and do not go into the technical aspects of specific technologies. Having described technology assessment, the chapter will then deal with portfolio and project evaluation with particular emphasis on externally sourced technology.

Technology assessment focuses on the age of a technology, its level of sophistication and its likely impact on the business over time. From an assessment of a technology we can decide if it is time to substitute a new technology. This will lead to decisions on technology acquisition and on the evaluation of specific technologies and projects. We then consider the effect of a number of technologies or a portfolio of projects. Finally, the chapter covers the subject of project evaluation and give details of specific techniques.

Classification of technologies

The main classifications of technology are shown in Table 5.1. Base technology is the lowest level and is also called enabling technology. This is the basic technology that enables a company to carry out its

Table 5.1 **Classification of technologies**

Base	Essential for the company to stay in business but competitors also have it
Key	Gives a competitive advantage, core competence of company
Emerging	Not yet proven, but if successful could be tomorrow's key technology

main business on a daily basis. This technology is not unique to the company and is similar to that used by its competitors. It gives no competitive advantage.

The next level up is called key technology, also called competitive technology. This is unique to the company and is what gives it its competitive advantage. Some firms in very competitive sectors do not have any key technology. They have the same base technology as their competitors. Other firms that are innovative have key technology that enables them to differentiate their products, to offer increased performance or lower prices, or to make better profits.

The highest level is called emerging technology. This technology is being introduced and has not yet proved itself. If successful, it will become the key technology of the firm in time. By that time the firm's competitors may have advanced, and what was previously key technology must then be considered base technology because competitors also have it.

Tramex Limited is a world leader in capacitative moisture meters. These meters use the fact that moisture changes the electrical capacitance of a substance, which affects its ability to transmit high frequency electrical signals. When Tramex started off in 1982 with a single instrument, its technology was unique and patented. This was its emerging technology. Its base technology was electronic assembly and its key technology was in aspects of electronic circuit design. After a few years capacitative measurement became its key technology and electronic circuit design became a base technology. In more recent times Tramex has developed digital meters, which have numerous advantages over analogue meters, and digital moisture measurement has become its emerging technology.

Technology assessment can be applied to new or old technologies, and companies can be evaluated by the type of technology they possess, which can say a lot about the type of firm. A useful classification of technology levels is shown in Table 5.2. Technologies are classified from base to emerging and the different levels suit organisations with different technological risk profiles.

Table 5.2 Technology classification and organisation types

Classification	Level	Organisation
Base technology	Low?	'Me too'
Key technology	Medium	'Follow the leader'
Emerging technology	High	'First to market'

The firm with only base technology could be described a 'Me too' and such firms are usually following a technology acquisition strategy of copying.

The firm with a medium level or key technology has a competitive advantage. Its technology and intellectual property give it a degree of product differentiation. It is not a technology leader but is trying hard to keep up. The organisation could be described as 'Follow the Leader'. It usually follows a technology acquisition strategy of modifying or improving, but it may also seek to keep up with the technology leader through acquiring technology by licensing or takeover.

The firm with the highest level has emerging technology. The organisation is usually 'First to Market' with a new product or service. It follows an inventing or improving technology acquisition strategy.

If an organisation seeks to change itself, say from 'Me too' to 'Follow the Leader', its technology acquisition strategy must move also. There can also be a mix of technologies in different divisions of the same company. By having a coherent corporate strategy, larger companies can use their technology strengths in one division to assist weaker divisions.

In the computer industry, Dell and Compaq could be regarded as 'Me too' companies as the manufacture of personal computers has become base technology in the computer industry. Note that base technology is not always 'low tech'. IBM is a follower, concentrating more on applications engineering than on research in recent years. Its technology is key technology but it now relies on others to do the fundamental research. Apple has consistently tried to be a leader but this has often led the company on a roller-coaster ride.

In 1997 Apple was facing bankruptcy. It was in Chapter 11, a provision in US bankruptcy law which allows companies to try to trade out of their financial problems. Apple had gone downhill since its founder Steve Jobs was let go. The company had lost its creative driver. But it had also been following a closed innovation strategy and this meant that few outsiders were able to develop applications for Apple machines. It had too many products and too little product differentiation. Steve

Jobs was recalled to save Apple and led the development of the iMac, which by May 2001 had become the best selling computer and had sold two million units in 12 months.

Technology life cycles

Another aspect of technology assessment is the age of the technology. Richard Foster[1] shows that technologies have life cycles just like products. Technology life cycles usually take the form of an S-curve. S-curves show the relationship between the performance of a technology and the R&D effort or investment over time. This can be seen in Figure 5.1. Initially a lot of money is spent on R&D for very little return. Gradually the technology starts to become successful and yields a considerable increase in technical (and commercial) performance. The curve rises steeply. Then it starts to approach the limit and diminishing returns set in. The curve levels off. Eventually, no further return can be obtained by investing in this technology and it is superseded by a new technology. A new S-curve starts.

The example illustrated in Figure 5.2 shows the development of audio hi-fi systems. In this case, the performance of vinyl records as a

Figure 5.1 An S-curve shows the relationship between technology performance and R&D effort

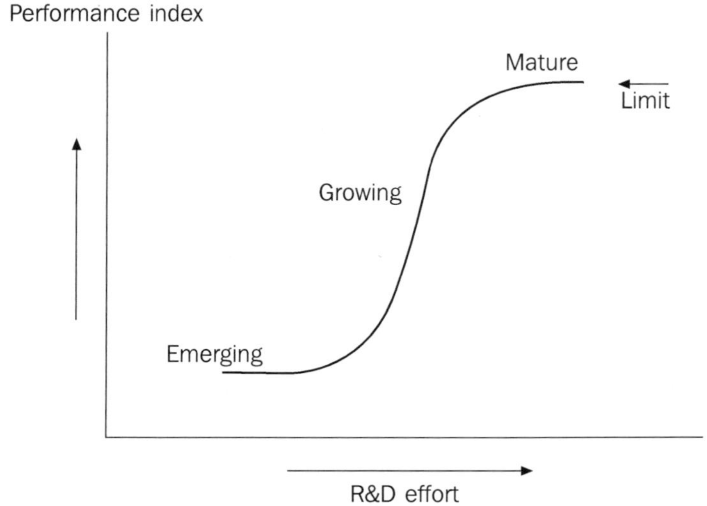

Figure 5.2 S-curve for Audio HiFi technology

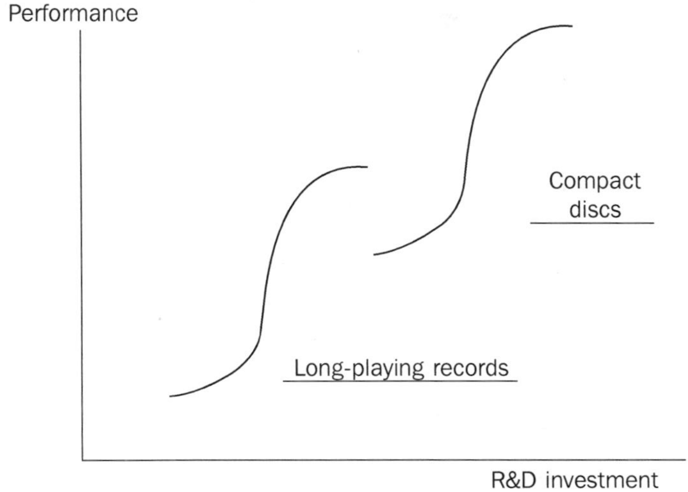

reproduction medium had begun to yield diminishing returns. The investment by Philips and Sony in the compact disc then began to pay dividends.

One could ask the question 'Does the extension of the use of CDs to digital data storage represent another S-curve?' The answer is that it is the same technology, but a new application has been found for it. This would have the effect of extending the curve upwards.

In carrying out its technology assessment, a company must ask itself where it is on its technology S-curve. If it is at the top of the curve, its technology has become base technology. The curve shows that further investment in that technology will yield diminishing returns. The company must be prepared to move to a new S-curve. This means it must be prepared to abandon its old technology and acquire new technology. If it cannot see how to invent new technical solutions to meet its business objectives, the other technology acquisition strategies such as licensing and purchase of technology should be considered.

Technology impact analysis

Technology impact analysis asks how a technology will impact on a company over time. It uses the principles of S-curves to enable a company to examine its current technologies to see where they are in their life cycles.

Technologies that are still on a rising part of the curve are key technologies. These are technologies that can continue to impact over the company's planning horizon. They will continue to make a contribution to the company's revenues in the short to medium term. They have not yet reached saturation level in terms of R&D investment.

Technologies that are far from the limit can still make a substantial impact, or delay the impact of superior substitutes.

If the S-curve is levelling off a company must consider technology substitution.

In 1967, National Cash Register (NCR) was the market leader in mechanical cash registers. Electronic cash registers had not appeared on the market although electronic computing had reached a point where the development of electronic cash registers was forecast. In 1971 the first electronic cash register appeared. Meanwhile NCR was doing electronic development but mainly as a support to its main product line rather than a substitute. In 1972 the market for cash registers was 10 per cent electronic and 90 per cent mechanical. By 1975 the ratio had reversed; NCR had only 10 per cent market share and had to write off $140m in obsolete stock and let 20 000 workers go.[2]

Stages in carrying out technology impact analysis

In carrying out an analysis one should draw the S-curve for each technology. It is not strictly necessary to draw the S-curve because in collecting the information to draw it, the impact of the technology becomes clear. However, the S-curve is the most revealing way to illustrate the age of a technology and if it has levelled off, the call for action is one that cannot be ignored. The stages on constructing a curve are:

1. Define performance parameters.
2. Calculate limits.
3. Plot historical part of curve.
4. Draw limit line at top of graph.
5. Project future part of curve.

Performance parameters are: product sales, market share, royalty income from licensing out, R&D expenditure and cost–benefit analysis of R&D expenditure, comparison with competitors and state of the art. In defining parameters it is best to concentrate on critical areas and produce a score of performance. This may be subjective and involve estimates. Some factors may be weighted to show their relative importance over other factors.

Indications of approaching maturity are easy to identify. Some of these are:

- Fully commercialised several years ago, many competitors, base technology.
- Decline in rate of change of technology or new applications.
- Decrease in patents or technical publications.
- Increased attempts by competitors to develop substitute technology.

Having obtained a measure of performance, the next step is to draw limits. One way of doing this is to get opinions on how the technology could be limited if any attribute was extended to its maximum. For example, if NCR looked at its mechanical cash register developments it might have seen that there was a limit to the speed of mechanical calculation, that it would be difficult to reduce weight and cost any further and that product reliability was unlikely to improve. The upper limit of performance is drawn as a horizontal line, which establishes the point at which the curve levels out.

Plotting the historical part of the curve is done by calculating or estimating the historical performance scores. These are then drawn on the graph, the upper limit is added and the gap represents the future progress of the technology towards maturity. The line that closes the gap can be drawn by mirroring the lower part of the curve, but this is a very crude approach. More sophisticated approaches use curve plotting software or attempts to predict the performance score some distance into the future.

Using the information obtained in plotting the S-curve, we can assess the impact of a technology, both technical and commercial. We can then assign scores for technical and commercial impact over time. For example the investigation may suggest that a particular technology will have a significant impact but not for 4 years. You then have to decide if you need to take action now or later.

Discontinuities

In constructing S-curves it is important to realise that you are only looking at a particular technology and trying to predict what will happen to it without knowing the impact of other emerging technologies. It is quite possible that a new technology may emerge while an earlier technology is still in its upward phase. This is called a discontinuity. The new technology is sometimes called a disruptive technology if its effect is to cut short the S-curve of the existing technology. The new technology can impact on the old by limiting the sales performance, but this is not always the case. The old technology still has scope for further development, and in many cases it may continue to compete. Technology impact analysis does not try to predict discontinuities, although technology forecasting, which is also used in technology impact analysis, may assist in predicting the arrival of a disruptive technology.

S-curves are relatively stable and their progression is predictable. Discontinuities occur when a variable goes from a stable to a chaotic region of variation and this is what happens when a company is jumping from one S-curve to the next. This is a time of great uncertainty and the transition has to be carefully managed.

Technology substitution

New technologies replace old ones because they provide a new opportunity to improve the return on R&D investment. If the opportunity is not grasped by a firm, it will be taken up by its competitors. The effect of the emergence of a substitute technology can be highly strategic, repositioning the firm for a new phase of growth or securing its future. The effect can be surprisingly sudden, and once it starts, its development can be quite regular and predictable. The technology has entered a more stable region of growth.

The response to technology maturity is to find a substitute technology. This is done by reviewing the possible technology acquisition strategies as described in Chapter 3. The company may decide to do R&D or if the new technology is beyond its present knowledge base it will have to acquire it externally. If no substitute is in sight, it is better to stay where you are but to be aware of the fact that investing in R&D is unlikely to produce benefits unless you are going to invest in the research phase to start you on a new S-curve.

Table 5.3 Examples of technology substitution

Transportation	Horses to automobiles Steam to diesel Rail to air Piston engines to jet engines
Materials	Natural fibres to synthetic Natural wood to MDF Metals to plastics Soap to detergents
Communications	Telegraph to telephone Telex to fax Mail to e-mail Landline to mobile

Possible substituting technologies can come from competing technologies that are already in the company, but not yet mainstream. These must be evaluated and brought forward if suitable. Another source is what is known as diffusing technologies. These are technologies that are already used in other businesses and that have potential for application in yours. Finally, emerging technologies must be examined to see if they can or should be acquired.

Conclusions from technology impact analysis

The impact of the firm's existing or current technologies can be assessed to see if they should be abandoned, maintained or developed, or if alternative technologies should be acquired externally. This depends on the availability of substitutes and their suitability. The estimated impact of substitutes over time will indicate whether these should be developed or acquired externally or just monitored until their impact becomes clearer. Table 5.4 shows some of the options based on the technology classification and its impact over time.

If a base technology is competing against a substitute it may be abandoned but it should be replaced by a key technology or even an emerging technology. If there is no substitute to the base technology, the company should at least maintain a competitive position by making sure its base technology is kept up to date.

An example of a base technology where there appears to be no substitute at present is electric power generation and distribution. While there

Table 5.4 Technology strategies based on time of impact

	Time of impact	Action
Current technologies		
Base (substitute in sight)	In the past	Abandon
Base (no substitute in sight)	In the past	Maintain competitive ability
Key	Present	Refine
Emerging	Impact in 0–3 years	Continue to develop
Substituting technologies		
Key	Impact in 0–3 years	Acquire immediately (may be too late to develop)
Key	Impact in 3–8 years	Invest in development or external acquisition
Emerging	Impact in 8+ years	Monitor progress

may be new energy sources in prospect, the basic principles of electricity generation, distribution and transmission have not changed in more than 50 years.

Portfolio management

Many companies deal with multiple technologies and projects. Decisions on technology acquisition are taken primarily for strategic reasons. Once technologies have been acquired, a decision-making process is also required to prioritise the development and commercialisation of technologies and to ensure that a balanced portfolio of projects, technologies and products is available to the company's divisions to enable the company to maintain its strategic advantage.

The purpose of carrying out a regular assessment of a project portfolio is:

- To make go/kill decisions on projects.
- To prioritise resources so that better projects get ahead quickly.
- To monitor the progress and effectiveness of the work on each project.
- To decide where a project is heading.
- To balance the portfolio.

The biggest problem in portfolio management is selecting between competing projects and the effect that this may have on the motivation of staff who have become attached to a particular programme of development. Killing off deserving but lower profile projects has always been one of the most difficult tasks in the management of a firm's technology and one that is often badly handled. Either the issue is ducked and a poor project survives to soak up valuable resources before eventually failing, or a project is killed off without considering the feelings and motivation of the staff who have been working on it.

Another issue in portfolio management is the impact of one technology on another. Technologies and projects compete for limited resources. To have a number of really exciting technologies or projects in development at the same time can be a major problem. If there are not enough resources, some of the projects are going to be downgraded and their potential may remain unfulfilled unless more resources are obtained or the technology can be is licensed out at an early stage.

Technology platforms

Increasingly, firms are seeking to acquire technology platforms, technological competencies or core technologies that give rise to a continuing range of products based on the platform technology. We saw examples of this with Honda's competence in small engines and 3M's competence in adhesives and coatings. This approach has become more noticeable in the software industry where technology platforms provide basic building blocks for a range of application products aimed at different sectors. Technology platforms are now a key component of many companies' technology portfolios.

Portfolio techniques

There are a number of elements that are key to good portfolio management. One is to have a good project evaluation system that gives scores to projects so that they can be ranked. It deals with early stage projects and helps with project selection and prioritisation. Using a scoring system causes poor projects to drop down the list and as the criteria applied are objective, it is more likely that the decision not to proceed will be accepted. Instead of feeling that their pet project has been torpedoed, staff are more likely to see it as failing to compete for resources with better projects. Project evaluation will be dealt with later in this chapter and in further detail in later chapters.

Projects can also be compared using the portfolio map shown in Figure 5.3. Here projects are evaluated on the basis of their probability of success and value to the company. Some projects may have a high probability of success but contribute only a small amount to the company's profits. These are classed as 'Bread and Butter' projects. Others that score on both fronts are classed as 'Pearls'. Those that score badly are classed as 'White Elephants' and those that have a high value but are a long shot are called 'Oysters'. Thus it can be seen that a company needs a few pearls and a good number of bread and butter projects. The occasional oyster may be worth the risk and the white elephants need to be culled.

As well as grouping the projects into the four quadrants, the projects can actually be plotted on this diagram in their relative positions and using circles to indicate their sizes. This is illustrated in Figure 5.4 where it can be seen that there is one really good pearl, one risky oyster and a large white elephant long past its sell-by date. This diagram is used to help get a good project mix.

Technology and project assessment

Technologies can be assessed by considering their potential impact over time as we saw above. Technologies can be either those currently owned

Figure 5.3 Portfolio mapping: taking a strategic view of the portfolio (PS = 'Probability of Success')

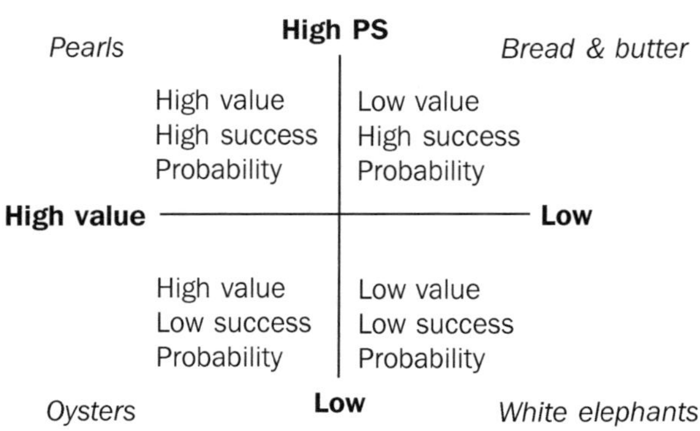

Figure 5.4 Portfolio mapping: getting a good project mix (PS = 'Probability of Success')

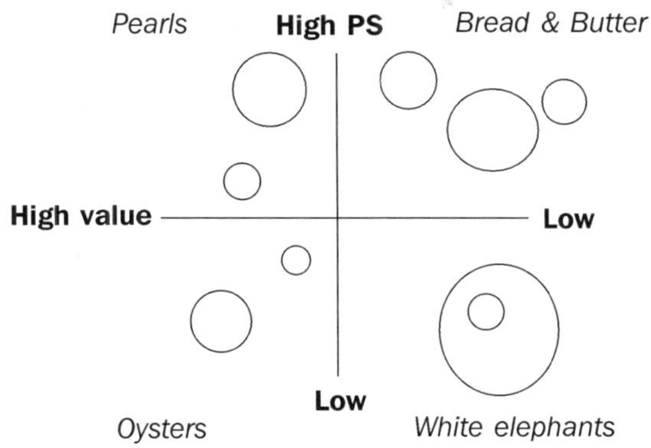

by the firm or potential substitutes. Projects are activities within the firm. They may be directed towards acquiring technologies, but project assessment is a method of evaluating competing activities rather than technologies.

In Chapter 1 we saw how research on success and failure in the introduction of new products demonstrated the importance of devoting sufficient resources to the evaluation of new opportunities, and how resources can be thrown away by allowing poorly researched or pet projects to proceed into the expensive stages of development and commercialisation.

Project evaluation

Many firms do poor project evaluation. There are many evaluation techniques and choosing the right one is difficult. Many of them depend on having reliable information that can be difficult to acquire when a project is at an early stage. Sometimes the information is available but the firm does not have the expertise to find it. Decisions are often taken on guesswork or hunches.

There are two kinds of evaluation errors: drop errors and go errors. Drop errors are where good projects are dropped because their potential was not revealed. Go errors are where potential failures get through the evaluation process. As 75 per cent of projects put into the expensive development phases have failed in the past, it is clear that errors in evaluation can have a huge impact on the success of a company and improving its project evaluation skills can have benefits that vastly outweigh the costs.

Cooper[3] reports that very little investment of time and effort goes into the front-end activities of project evaluation. Market evaluations are rarely carried out in any depth, 57 per cent of companies surveyed failing to carry out an in-depth market survey. Decisions on whether to proceed with a project are often taken by one person rather than by a multidisciplinary team and half of the development budget is spent on projects that fail or are discontinued.

The subject of review meetings that set 'Must Meet' criteria was introduced in Chapter 3 in the context of reviewing ideas resulting from a product or technology search. A multidisciplinary team is set up consisting of representatives from R&D, marketing, finance and other departments. This 'innovation committee' can perform a number of functions. It can assess ideas from suggestion schemes, review new product or process proposals, assess product and technology search findings and decide on allowing projects to progress to further stages and allocate resources. To do these things effectively it needs to set objective criteria for decision making that can then be applied and should be rigidly adhered to.

Table 5.5 shows a simplified set of 'Must Meet' criteria. This is used for initial screening, the first filter for ideas or project proposals. If any idea or proposal fails to meet any of the criteria, it is killed immediately. However, sometimes a breakthrough idea comes up that does not meet the criteria. It might be a new technology that requires investment in new equipment, the current criteria specifying that all new products had to be capable of manufacture on existing equipment. It is always good

Table 5.5 Must meet criteria for project evaluation

Must meet criteria	Idea 1	Idea 2	Idea 3	Idea 4
Market need/demand				
Meets profitability level				
Fits business strategy				
Fits technology strategy				
Unique selling points				

☑ = OK.
☒ = rejected.
Projects must meet all criteria. Failure to meet any criterion results in rejection.

to treat these breakthrough ideas as exceptions and subject them to a separate detailed evaluation according to relevant criteria.

The next stage of evaluation is 'Should Meet' criteria. Here, projects that have passed the 'Must Meet' conditions of the first filter are subjected to a more detailed evaluation. This is a scoring system and enables projects to be judged on a wider range of criteria and ranked. The lower scoring projects effectively reject themselves thus reducing the possibility of pet projects or staff dissatisfaction. In sporting terms, the first filter is a knockout system but the second filter is a league table. A much simplified set of 'Should Meet' criteria is shown in Table 5.6. While a small number of factors may be considered for 'Must Meet', a much larger number need to be considered for 'Should Meet' and adequate time should be devoted to obtaining consensus on the criteria before any projects are evaluated. Once the criteria are agreed it is worthwhile to trial them on previous projects. This often shows that some previous projects should not have gone ahead. It also tests whether the criteria are realistic. In cases where some criteria are considered to be more important than others in the decision-making process, weighting factors can be applied.

Table 5.6 **Should meet criteria for project evaluation**

Should meet criteria	Idea 1	Idea 2	Idea 3	Idea 4
Environmental issues				
Manufacturing/production issues				
Intellectual property issues				
Packaging logistics localisation				
Distribution				
QC/standards				
Training				
Sales & marketing				
Financial issues (funding, ROI)				
R&D issues				
Total score				

Score each idea on a scale of 1–10 on the issues in the left-hand column.

Patent searching in technology and project assessment

Patent searching is an important tool in technology assessment and project evaluation. It enables you to find answers to the following questions:

- Is the technology new?
- Who owns the technology?
- Is there other technology that is similar?
- How well is the technology protected?

The subject of patent searching was introduced in Chapter 3 as a method of finding new technologies and licences. We will leave the details of patent search techniques until after the subject of patents has been explained in later chapters. Patent documents contain details of previous patents found in patent office searches as part of the process of patent examination. These are called patent citations. The evolution of a particular technology can be tracked by carrying out a search and then picking leading patents of leading companies and digging down through the references cited against them. In this way, earlier patents showing milestones in the development of the technology are uncovered. It is also possible to compare the number of patents owned by your company and its competitors in certain key technology areas.[3]

Patent databases can be searched under specific patent technology index classes and dates. This can readily reveal statistics on patent filings showing the number of filings in a technology area on an annual basis. Technology life-cycle information can be gleaned from this type of analysis.

In looking at how well a technology is protected, you can find the territorial coverage of patents, the age of patents and the scope of patent claims. These points will be covered later in Chapter 9 where we consider the assessment of new projects and technologies, both yours and your competitors'.

Notes

1. Foster, R.N. (1987) *Innovation: The Attacker's Advantage*. Pan Books.

2. Ibid. pages 139–141.
3. Cooper, R.G. (2001) *Winning at New Products*. Perseus, pp. 22–49.
4. Germeraad, P. (1999) Intellectual property in a time of change. *Research Technology Management* 42 (6): 34–39.

6

Market-led R&D project management

Introduction

R&D projects are managed by breaking them down into stages. Each stage requires a different approach. In this chapter we will follow the innovation stages described in Chapter 1. These are: idea – evaluation – development – market. All R&D projects go through these stages but for the sake of simplicity we will refer in this chapter to product rather than process or technology development. The product development stages were expanded in Chapter 1 to cover the additional stages of concept evaluation, prototype development, development for production, testing and launch. We also saw that for engineering products the process stages are slightly different to that for chemical and food products.

Recent thinking in R&D management has focused on the integration of marketing criteria into the project management stages. This has the purpose of ensuring that the market opportunity is correctly identified and is not lost sight of as the project progresses through development. Frequently scientists and engineers engaged in development have been accused of adding cost and functions resulting in the product being late, too complex and over budget. These issues are central to this chapter.

Stage Gate™

The Stage Gate™ project management technique has been promoted by Robert Cooper and is well described in his book *Winning at New Products.*[1] This technique breaks each project into stages but between each stage there is a 'gate', a decision point which cannot be passed unless preset evaluation criteria have been fulfilled. The basic Stage Gate™ decision making process is shown in Figure 6.1.

The first stage is usually the idea generation stage. This is followed by one or more evaluation stages consisting of an initial filtering as described in the previous chapter and then business analysis. Business analysis covers detailed market and financial evaluation and investment issues if new equipment is required. The project then enters the development phase that could include a number of stages leading eventually to the final stages of testing and product launch. Between each stage then is a decision gate. As in must and should meet criteria, these filter points provide objective pre-agreed standards that the project must attain. Projects cannot move to the next stage unless they meet the gate criteria.

Criteria for the stages after evaluation depend on the nature of each project. For an engineering product it is usual to go to prototype development and test followed by development for production. This is an important decision often involving significant investment in tooling for mass production. Clearly, criteria will be set for the performance of the prototype, before the major tooling investment can be agreed.

A slightly more developed version of Stage Gate is shown in Figure 6.2. This illustrates the decision-making process in more detail. In the case of projects that have entered the development phase, there are three possible outcomes, go, kill and redo. Projects that do not meet the gate criteria may be sent back to repeat the stage if it is felt that there is scope for meeting the targets.

Product specifications

It was stated in Chapter 4, in the context of idea generation, that a product specification is a definition of the requirements that a product

Figure 6.1 Typical Stage Gate™ project stages

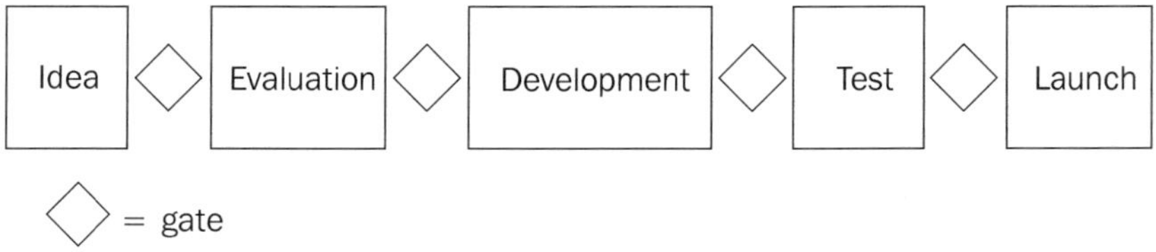

Figure 6.2 The Decision-making process of Stage-Gate™

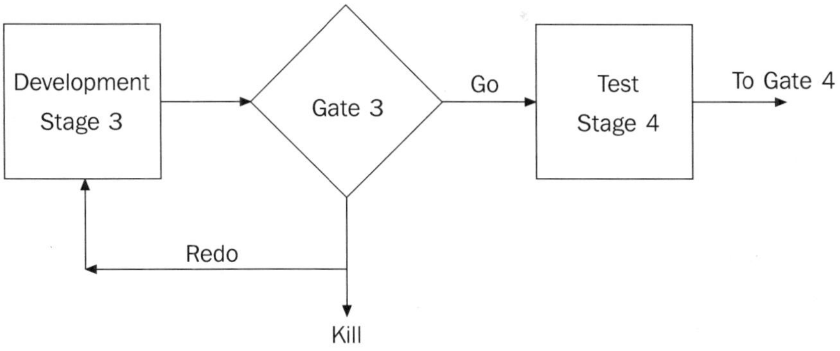

should meet without defining how they are met. In other words, a product specification does not contain a technical solution. If we think of the peristaltic actuator described in Chapter 4 we could state a product specification for is as follows: 'A long stroke actuator with a stroke in excess of thirty metres capable of opening the doors of an aircraft hangar.'

Note that we have not said that it should be pneumatic or that it should be a cylinder. So a correct approach to product specifications focuses on the customer, on the market opportunity not on the technical solution. Engineers and scientists often latch on to the most obvious technical solution and pursue it regardless, thereby failing to consider the possibility of a creative solution or a new and more suitable technology.

The product specification is central to the achieving of the end result, a unique differentiated product that gives perceived benefits to the consumer. There are three major market-driven elements of R&D project management. If these are respected, the chances of success are greatly enhanced. These are:

- correctly identifying the product specification
- keeping to the product specification as the project proceeds
- negotiating changes in the product specification if required.

Correctly identifying the product specification

A good market knowledge is required to correctly identify the requirements that the new product should meet. This means knowing the difference between customer needs and wants. Customer needs are their underlying unspoken requirements, which are not elicited by direct questions. Customer wants are the unimaginative answers to poorly structured market research questions. Henry Ford is said to have stated 'if I'd asked people what they want I would have made a faster horse'. He understood that they needed something more than a horse because he understood the customers' motivations and unspoken needs.

It is said that in the area of consumer marketing or business to consumer (B2C) the decision to purchase is 20 per cent rational/80 per cent emotional. The ratio is said to be the opposite for business to business (B2B) purchasing decisions, which are mainly based on price, performance and other rational criteria. The type of market research needed to identify customers' needs is called diagnostic research. Instead of asking the customer what they want, the researcher asks more general questions about the customer's preferences often without describing a particular product concept.

If Akio Morita, the inventor of the Walkman had been prevailed upon to do diagnostic research we could imagine that he might have asked potential customers if they would like to listen to music on the bus or the train. He might have asked if they would prefer to listen to their own choice of music or to whatever was on the radio. He might have used focus groups. These are groups of selected consumers who are asked a series of questions or shown product samples and asked for their reactions. While their answers are important, researchers watching them through one-way mirrors are noting their expressions and body language, often gaining more insight from unconscious signals than from the questionnaires. The more the product relates to ideas of esteem rather than practical use, the more important are the unspoken needs. Thus diagnostic research is used by car manufacturers as well as by manufacturers of cosmetics.

Diagnostic research is often used as input to idea generation phase and customers are often recruited as part of a brainstorming session. This is sometimes called 'customer stimulus'. Once the customer's needs have been identified the role of the customer in the stage review process is often taken by a member of the marketing department who is often called the 'internal client'. He or she then becomes responsible for making sure that the product specification is respected in setting the criteria for the gates and in subsequent project review meetings.

Keeping to the product specification as the project proceeds

Many R&D projects suffer from what is called 'function creep'. It can be too easy to add functions especially in software projects. Studies by the Standish Group[2] show that 31 per cent of software projects were cancelled before completion and that the main reasons for failure were lack of user input and incomplete or changing project specifications. In Florida the mandated child welfare system project was started in 1990 with a budget of $32 million with a staff of 100 and a target completion date of 1998. At last report the budget was $170 million with a target completion date of 2005. However, a similar project with much clearer specifications and better management was carried out in Minnesota in one year using eight people and a budget of $1.1 million.

In Ireland the PPARS project was set up in 1995 to provide a new payroll and human resource management system for the 130 000 staff in the health service. It was suspended in October 2005 having escalated in cost from €9 million to €160 million. By 2005 is was reported that it could only cover one-third of the staff and the estimated cost at completion would be €231 million. Changes had been made without consulting the hospitals and without overall co-ordination. A report into the project found that a lack of leadership had allowed the cost to spiral out of control. As one politician saw it 'Everyone is responsible and no one is accountable.'[3]

Another report by the Standish Group showed that only 45 per cent of the features were ever used in software programs and only 20 per cent were used often.[4] While 'feature-itis' is clearly a problem in software it is also found in mobile phones and in many other products.

'Customer insight'[5] is the term used for preventing the development from departing from the product specification. As the project proceeds the product features are checked with potential customers to see if they meet the customers' needs. If they do, it is then possible to predict that the customers' needs are likely to be met by the finished product. In many cases companies wait until the market testing phase and this is often far too late to make changes.

As well as checking adherence to the customers needs it is also advisable to allow for the identification and inclusion of features that have a perceived value to the customer. Charles Miller and David Swaddling define Customer-Perceived Value in *The PDMA Toolbook for New Product Development*.[6] as follows: 'Consumer-perceived value is the result of the customer's evaluation of all the benefits and all the costs of an offering as compared to that customer's perceived alternatives. It's the basis on which customers decide to buy things'.

Quick Metal, an example of customer-perceived value

660 Quick Metal® Retaining Compound is a creamy, non-running gel that is applied to cylindrical parts of machines to fill surface imperfections and repair worn areas. It is manufactured by Loctite Corporation and contains no metal, but hardens to metal strength.

It was originally launched as RC-601 and as it was a new concept in machinery repair, it was priced at only $10 per tube to gain market acceptance. It was aimed at production engineers and was backed up by detailed technical specifications. The product failed to take off and after a year it was recalled from sale and the value of the unsold stock was refunded to distributors at considerable cost to Loctite.[7]

On investigation of the project it was found that the product was aimed at the wrong market. Production engineers are slow to approve new techniques and were suspicious of the low price. It should have been promoted to maintenance fitters and was subsequently relaunched in a simpler form with a catchy name and less technical details. The price was raised to $20 to reflect its value and it became a huge success.

Negotiating changes in the product specification if required

In emphasising the importance of keeping to the product specification it should not be assumed that the product specification is cast in stone at the start of the project. It is the responsibility of the internal customer to review the customer needs and to make sure that changes in those needs are recognised or that compromises in meeting those needs are negotiated and agreed.

For example it may emerge in the course of development that the target price cannot be met. In many projects what happens then is that the product is launched at a higher price and sales are less than originally forecast. However, if the correct steps are taken, it may be possible to negotiate a change in performance to match the target price. By negotiation between the members of the project team it may be agreed to go back to the customer with a reduced specification, or if this cannot be achieved, to go back with the higher price and see if it is justified on the basis of customer-perceived value.

This method of managing R&D projects is called 'Agile' and has come from the software industry. In software development it has proved relatively easy to refer back to the customer and negotiate changes in specifications as the project proceeds. A statement by the Agile Alliance[8] describes a much more flexible method of managing a software development project:

> We are uncovering better ways of developing software by doing it and helping others do it. Through this work we have come to value:
>
> Individuals and interactions over processes and tools
> Working software over comprehensive documentation
> Customer collaboration over contract negotiation
> Responding to change over following a plan.

Clearly, as many software projects are developed directly for the end-user it is easier to refer back and negotiate changes than it would be in the case of developing a non-software product at some remove from a range of potential customers. It is also easier to change software than it is to change physical products. Nevertheless, the principles of Agile product development are increasingly being adopted by development teams in other sectors.

One of the principles of Agile development is the introduction of a 'customer-developer interface',[9] which is used to set up a structure for the negotiation of changes in the product specification. Another principle is to use good design to achieve simplicity. By rigorous prioritisation of features early in the project, feature-itis can be eliminated. Finally, by building in an adaptive iteration cycle, rapid feedback is obtained and changes are made quickly as the project proceeds.

The final stages

Confirmatory market research is then carried out at the market test phase. By this stage there should be reasonable confidence that the product specifications are being met and the purpose of confirmatory market research is as it says, to confirm that the product meets the customer's needs. This type of market research follows more traditional lines. The customer is shown product samples and asked for comments. The customer response can be based more on rational judgement than in the earlier phases such as focus groups.

The principles of market-led project management are summarised in Figure 6.3. This diagram based on a combination of the principles of involving customers promoted by Innovaro and on Agile promoted by the Cutter Consortium[10] shows how projects can be managed using Stage Gate™ yet ensuring that what emerges for the process is tailored more closely to the customer's requirements. As the project proceeds the risk factors are steadily reduced thus giving the best chance of a successful outcome.

A final word on Stage Gate™

Stage Gate™ is not ideal for very innovative products. They tend to get rejected, often because the criteria have been set for more ordinary projects. Projects that are considered to have exciting potential may need to be considered separately and under different criteria. Stage Gate™ is good for ordinary new products and product improvements where process is reasonably predictable.

It is always worthwhile to carry out a post-implementation review of each project. This can reveal what worked well and what did not work well. By reviewing a project objectively the findings can be fed back into the development process to improve performance on future projects. In one company the marketing department tended to give very optimistic sales forecasts and the design team tended to be blamed for high production

Figure 6.3 Stage Gate™ Process for market-led project management

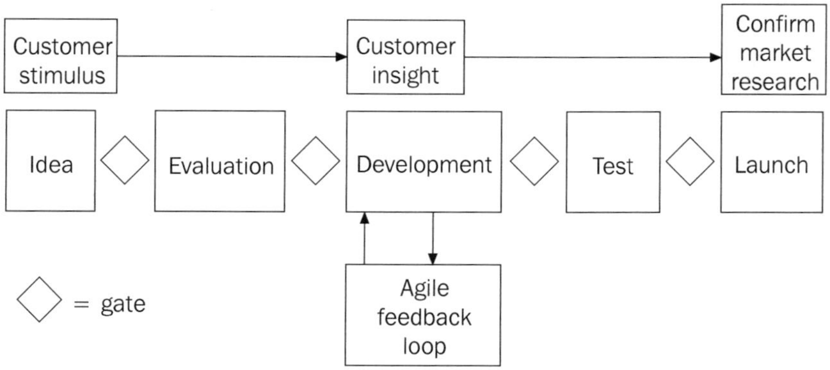

costs. However, when the project was reviewed it was shown that the costs were high because the forecast volumes were not achieved.

Notes

1. Cooper, R.G. (2001) *Winning at New Products*. Perseus, pp. 130–141.
2. See *http://www.standishgroup.com/sample_research/chaos_1994_2.php* (accessed 22/03/06).
3. See *Irish Times* article accessed on 22/03/06 at *http://www.ireland.com/newspaper/front/2006/0304/21021859HM1PPARS.html*
4. Standish Group op. cit.
5. See *http://www.innovaro.com/* accessed on 30/03/06
6. Miller, C. and Swaddling D.C. (2002) *The PDMA Toolbook for New Product Development*. New York: Wiley, pp. 87–114.
7. Best, R.J. (1997) *Market-based Management: Strategies for Growing Customer Value and Profitability*. Prentice Hall. As quoted in Miller and Swaddling op. cit. p. 87.
8. See *http://www.agilealliance.org* accessed on 03/04/06.
9. Highsmith, J. Cutter Consortium, *http://www.cutter.com*
10. See Innovaro op. cit and Cutter op. cit.

7

Intellectual property: patents

A knowledge of intellectual property (IP) is essential in the implementation of any technology acquisition strategy. How can you invent or improve a product without knowing what can be protected and what is already protected? How can you copy without the risk of being sued? How can you license-in technology without the risk of paying for something that might be already available free of charge? Understanding IP enables a firm to protect its new technology and to assess the protection of any externally identified technology.

This chapter deals with patents, the most important type of protection for new technology. Chapter 8 deals with copyright, designs, trade marks and other forms of protection. The legal treatment in this and succeeding units will be primarily based on English law with references to other countries where appropriate.

Introduction to intellectual property

IP means the property of the intellect. It is a property that can be built up and extended like a factory, but it is not a physical property of bricks and mortar or a fixed asset like a building. It is considered to be an intangible asset, one that cannot be touched physically but exists in the form of information and legal rights to control the use of that information. It can be bought and sold like any other form of property. It can also be rented out by the process called licensing.

IP is sometimes called IPR or intellectual property rights. The word 'rights' refers to the legal aspect of IP. A patent is a monopoly right, as is a registered design or a trade mark. This means that it belongs initially to the person or persons or organisation that created it and they can control the rights to exploit it. Other rights exist that are not monopoly rights as will be explained later. An example is reproduction rights such as in copyright.

Recent developments in markets, technology and in the legal and international business environment have pinpointed the importance of IP. These are:

- Increased competition due to freer trade and globalisation.
- The need to protect investment in technology and R&D.
- Acceptance of the concept of 'intellectual capital'.
- The increased use of licensing.
- Freer dissemination of information.

We are now in the era of the Single European Market. In the EU, there are no longer any customs barriers, and the so-called technical barriers to trade, such as standards, are in the process of being harmonised. It is no longer possible to carve up the EU into separate countries for trade purposes, so technology developers now use a different approach. Instead of protecting themselves territorially, they protect themselves technologically. Patents, trade marks and copyrights still provide the same protection as they always did and are increasingly used where the former territorial or national means of market protection are no longer available. With increasing globalisation many products are now made in the Far East. European companies who subcontract their manufacturing to lower cost economies now make their profits on trading are rather than on manufacturing. The owner of the IP reaps the profit regardless of where the product is made.

Product and process development has now become so expensive that developers are reluctant to invest in an expensive development programme without knowing that they can protect themselves against copying. If a competitor copies a product they can place the copied product on the market without having to recover the R&D investment. This gives them an advantage over the company that has invested large sums in R&D. Consequently, R&D investors must ensure that they have protected their investment from copying.

We have reached a stage of industrial sophistication where the product of the brain is more important than the product of the hand. Nowadays, most of the real strength of many corporations lies in the expertise of their staff, their intellectual assets or intellectual capital as it has come to be called. When companies are sold, it is usually for a sum many times higher than the asset value on the balance sheet. For high tech companies it can be over 10 times higher. IP is a component of the intellectual capital of a company.

Increasingly, licensing is used as a means for transferring technology from a technology developing organisation to a manufacturing and marketing organisation. Universities, research institutes and R&D companies operate this way. These organisations must invest in protection for their technology because without IP, there is no means of obtaining royalty payments for licences. If there is no protection, the technology will be copied.

With global communications and the ease of transfer of information over the Internet, information on new technology can be quickly disseminated, and it is very easy to lose control over the level of dissemination. Companies can benefit by knowing how to make information on their developments widely available to potential customers or licensees; however, they must be careful in releasing information that can affect their IP rights or competitive position.

Forms of intellectual property

IP can take a number of forms. These differ widely in their method of protection and in the types of products and technology that they can protect.

- Patents – for inventions.
- Copyrights – for literary and artistic works and software.
- Design rights – for industrial designs.
- Trade marks – for distinctive names, logos or graphics.
- Know-how – for secret information of commercial value.

Patents are used to protect inventions, i.e. new products or processes containing a new principle or idea. The inventions must be capable of industrial application (i.e. useful).

Copyright, while originally intended to protect the authors of literary, artistic and dramatic works, has now been extended to include most forms of reproduction of original works and can also cover products such as software.

Design rights may be registered or unregistered. Registered designs protect the aesthetic features of a product, for example its shape or styling. The unregistered type of design right is a form of industrial copyright in three dimensions.

Trade marks are distinctive names, logos or graphic representations for products or services and are used to make a connection between the promotion of the goods or services and the supplier. They can acquire considerable importance because of the goodwill they protect.

Know-how is secret or proprietary information of commercial value. It is not protected by any of the legal systems above, but by being kept secret.

Patents, registered designs and trade marks require to be registered in Patent Offices. Copyrights, unregistered design rights and know-how do not require any registration and therefore do not have any costs associated with obtaining them.

Intellectual property law internationally

There are separate laws in each country governing patents, copyright, designs and trade marks. For this reason these are called statutory systems. The World Intellectual Property Organisation (WIPO) works to ensure cooperation between countries and to harmonise their IP laws. Most European countries have now entered the European Patent System and operate according to the same patent laws. This system is independent of the EU and covers countries such as Switzerland and Turkey, which are not in the EU.

Harmonisation in other areas lags behind, although the European community has introduced a Community Trade Mark and a Community Registered Design. There is a proposal for a Community Patent but the success of the European Patent System has made the need for a community-wide patent less urgent. New directives on designs and software have gone a long way to harmonise EU law in these areas.

US law has developed independently of European law and there are many differences. The most important differences in patent law will be referred to below. However, recent changes in US law are bringing it more into line with the rest of the world.

Patents

A patent is a monopoly granted by a state to the first inventor of a new invention in return for a full disclosure of the invention.

A monopoly is a right of exclusive use that is usually granted by a government. A patent gives the owner of it exclusive control over its exploitation. Each country grants its own patent rights covering its own territory, and a patent in one country has no effect in another country.

An invention is a new product or process, something that contains a new principle or an improvement on a previous invention.

It is essential to be the first person to invent a particular invention in order to obtain valid patent protection and the invention must be new.

Each state makes a bargain with the inventor or inventors whereby, in turn for fully disclosing the details of the invention, the state will grant a patent that gives monopoly rights to the owner of the invention for a period usually of 20 years.

The full disclosure of the invention enables the patent office in each country to assess whether the invention qualifies for the grant of a patent. It also helps to promote the publication of details of new technology so that when the monopoly period has expired, the invention can be freely used by others. Because of the requirement for full disclosure, the patent system is the biggest and best indexed source of technical information in the world with over 1 million patent applications being filed each year.

Novelty

To qualify for patent protection an invention must be new and inventive and capable of industrial application. Newness in patent law is described as 'novelty', and the concept of novelty can be considered to be the same idea as secrecy. Disclosing an invention to people or publicising it before applying for a patent can destroy its novelty and prevent a valid patent being obtained. Disclosure means disclosure to the public or telling people about the invention without issuing a warning that the information is confidential. In patent law, the word 'novel' means 'unknown to the public'. Therefore the making of the disclosure destroys the novelty of the invention, and if the invention is no longer novel at the time of applying for a patent, a valid patent will not be obtained.

Inventive step

The invention must also be inventive. This means that is must not be obvious to a person 'skilled in the art' if presented with the details of the problem requiring solution. Some element of inspiration is therefore required in arriving at the invention and mere design improvements are not considered patentable. Inventive step is difficult to assess in some cases and arguments in favour of inventiveness are sometimes best made with hindsight.

Industrial application

The invention must also be capable of industrial application. A surgeon could invent a new surgical procedure or a mathematician could prove a new theorem but these are not considered patentable because they are not applied industrially. However, software is increasingly being found to be patentable, especially where it has a hardware element.

The other requirement for patentability is that an invention should not be contrary to a country's law or public morality.

Patent applications

The inventor(s) of a new invention can apply for a patent, but increasingly these days, patents are being applied for by companies because most development work now requires resources beyond the means of individuals. If an inventor is an employee, the invention usually belongs to the employer.

Before applying for a patent, a specification is drawn up describing the invention. It is important that the specification should be correctly and accurately drawn up by a patent agent as there are requirements in the structure of a patent specification that are not obvious to the layperson or even to the technical professional. Mistakes at this early stage can be expensive. It is important to define correctly what the invention is in patenting terms and to make sure that the specification fully describes the invention in the broadest possible terms. This specification will be used subsequently, to prove that the inventor was the first person internationally to make this particular invention.

First filing

Although you may want to protect your invention in a number of countries or even worldwide, it is only necessary to file an application initially in one country, usually your own. There is an international convention by which countries recognise the date of first filing, which is called the priority date. This date is recognised by other countries provided an application is made in those countries within 12 months of the first filing. This means that you could invent something in the UK and file an application in the UK today, while a person in Germany might invent and file a patent application for the same invention tomorrow. Although the person in Germany filed in Germany ahead of you, you would still be entitled to a German patent and they would not.

The initial application only establishes a priority date and does not give any legal protection. You must wait until your patent has been granted before you can enforce your rights.

Patent claims

Patent claims define the scope of the monopoly claimed by the applicant. This is like laying claim to a territory of technology. The operation of the patent examination system can be thought of in terms of making sure there are no overlaps between what the applicant claims as new and what has already been done by others. Consequently, the strategy and legal wording involved in drafting claims is very important and this is one of the key skills of a patent agent. As well as drafting specifications and claims, patent agents provide access to the international patent system through a network of colleagues in other countries, thereby enabling an inventor in one country to obtain patent protection throughout the world.

The strategy of patent claiming is to identify what is new and advantageous about the invention, reduce this to legal, broad, but precise wording and then construct a series of statements that describe the invention in a number of different ways with narrower alternatives, so that if one claim is rejected by a patent office or a court of law, another claim may be successful.

Patent examination

The patent examination process consists of the following steps:

- Submitting the specification with claims.
- Carrying out a search.
- Publishing the specification and search report.
- Examining citations and filing response.
- Acceptance or rejection.
- Grant.
- (Opposition, revocation).

The patent specification, complete with a set of claims, is submitted for examination usually at the end of the 12-month period from the date of filing, but examination can be requested earlier if required. The examining office carries out a search based on the claims of the patent specification. The search usually goes back about 50 years through published patents and technical literature of the major industrial countries. If the claims of the patent specification are too broad, a large number of patent citations will be attracted that will cause considerable difficulty for the applicant as objections in relation to each of these will have to be answered. The patent citations are examined in conjunction with the patent agent and a response is prepared explaining why the claimed invention is different from the prior art. This response may result in a narrowing down of the claims originally filed and the patent agent at this stage attempts to get a set of claims that are wide enough to give commercial protection to the applicant. If this tactic is successful, the patent is accepted and prepared for grant subject to the revised claims.

In Europe, a patent application is published with a search report 6 months after the date of filing the request for examination. This means that the applicant's patent details are now in the public domain and the invention is no longer novel in patenting terms. The novelty can be destroyed earlier by the patent applicant publishing details of the invention or putting a product on the market. Once the novelty is destroyed, it is no longer possible to delay the patent programme by refiling the patent application. Similarly, it is not possible for a person to see an invention on the market in one country and to validly patent it in another country. The person would not be the first inventor and the invention would not be novel.

In Europe, after a patent is granted it is open to opposition or revocation proceedings by any person or organisation that feels that it can give good cause why the patent should not have been granted. Once the patent specification has been published, potential competitors can find out the patenting intentions of the applicant. Large companies make it their business to study each other's published patent applications and to enter opposition proceedings where they feel that they have a case. Patents can also be revoked if they are found subsequently to be invalid usually as a result of a patent infringement action in court.

International filing options

To have the initial priority date recognised, applications must be filed in other countries within 12 months of the initial filing date. Most companies or organisations will want to ensure the widest geographical protection for their invention. Even if they cannot market it worldwide they may want to license it in countries where they are not active at present. This can produce problems in terms of cost as the patent system is very expensive and filings in non-English speaking countries require expensive technical/legal translations.

As well as translations, some of the cost arises from the fact that different countries are doing the same patent examination in parallel with each other. This multiplication of effort and cost was addressed by the World Intellectual Property Organisation by setting up a number of options to reduce duplication of effort and postpone translation costs. These are the Patent Cooperation Treaty and the European Patent Office.

Patent Cooperation Treaty

The Patent Cooperation Treaty (PCT) is a international patent filing system that is of great benefit to organisations that need time to develop and exploit their inventions. It postpones certain costs and decisions for up to 30 months from the date of originally applying (the priority date), and this means that an applicant organisation can apply and keep its options open to obtain patents in over 120 countries and then finally decide on the countries selected after detailed commercial information is available. The PCT timetable is shown in Figure 7.1.

Figure 7.1 Patent Cooperation Treaty filing system timetable

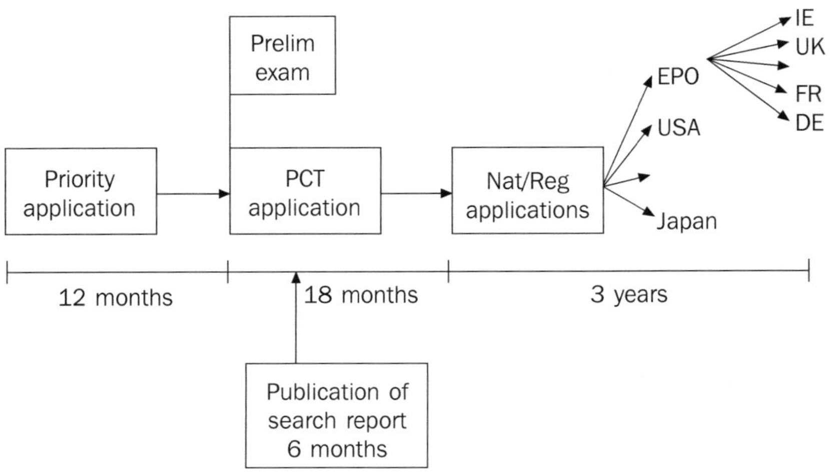

Twelve months after making the priority application, an application with a set of claims is made through the PCT system, which is accessed through national patent offices. The application is sent to an examining office and a search carried out. Applications can be made in English, French, Spanish, German, Russian, Chinese or Japanese and the office will deal with the applicant in the language chosen. The application is then published after 6 months, usually with a search report. The patent specification is now in the public domain and you can look it up on the Internet. An optional preliminary examination can also be carried out and an examination report sent to the applicant. This is merely a dry run through the patent system and can be used by the applicant to assess the chances of patent grant or to refine patent claims so as to have an easier passage through the next phase, examination in a national system such as the USA or a regional system such as the European Patent Office.

As can be seen, the PCT system slows down the entry to the final or substantive examination process and it can take 4 years or more before granted patents emerge from this route. However, it has the advantage that the applicant can keep options open for up to 30 months from the date of filing before finalising the countries in which the investment in patent protection will be made.

European Patent Office

The European Patent Office opened in Munich in 1981 to provide one central examination for European countries and to attempt to reduce the costs and problems involved in obtaining patents in Europe where there are a large number of small countries speaking many different languages. The avoidance of duplication and translation costs has greatly facilitated the granting of patents in Europe. Applications can be made in either of three languages, English, French or German and the Office will respond to the applicant in the chosen language. As in the PCT system, the application is published 6 months after the date of applying and usually with search report.

When a European patent is granted, it has no force on its own, but it acts as an instruction to the national patent offices to grant a national patent in each country. These patents are the same as national patents, which would have been examined in the country in question and apply in that country only. This process of converting to national patents can be expensive as translations into each national language are required so that each citizen in his or her own country can know what they may not copy. The European Patent Office is not an EU institution and does not grant a community-wide patent. Such a community-wide patent would apply as if all countries of the EU were a single country.

US patents

There are a number of differences between European and US patents. In the USA they recognise the date of inventing, and the inventor has 1 year from this date in which to validly apply for a patent. In Europe it is the first to file who gets the patent. For this reason US research organisations are very disciplined about writing up laboratory notebooks and getting them signed and dated at the end of each day. This practice is also recommended to non-US companies that intend to file in the US as they may need to prove that they made an invention before the date of a subsequent US inventor. Also in the USA it is always the inventor who applies and then assigns the patent to the employer. For this reason inventors are always named on US patent specifications but this is not a requirement in Europe where patents are often filed in the name of companies.

In the past, US patent specifications were not published until the patent was granted. However, the US patent system is gradually coming closer to European and international practice and since March 2001 US patent applications have been published. These are published 18 months after the date of filing.

Patents filed in the US before 8 June 1995 lasted for 17 years from the date of grant. In Europe, it is 20 years from the date of filing the specification with claims. US applications filed after 8 June 1995 will have a 20-year term from the date of filing or 17 years from the date of grant, whichever is the longer.

Patenting of genes

In general the identification of genes is regarded as a discovery and is not considered to be patentable. A patent cannot cover a gene in its natural surroundings. It is only patentable if produced outside its natural surroundings or used to make an industrially applicable product such as a drug. Industrial application refers to the manufacture of something with a definite commercial use and prevents patenting of genes for speculative purposes before an application has been identified.

Short-term patents

Some countries such as Ireland, Belgium and Germany have short-term or 'petty patent' systems. These are used for protecting inventions with a lower inventive content and provide a more limited form of protection for a shorter period usually between 6 and 10 years.

The short-term patent works on a different principle to the normal patent examination process. Short-term patents are granted automatically without examination but may not necessarily be valid. If infringement takes place, the court may have to decide on the validity of the patent before deciding whether it has been infringed. Before drawing the existence of its short-term patent to the attention of a possible infringer, the applicant must have a search carried out in a patent office to prove the novelty of the invention.

Infringement of patent rights

A patent covers:

- Making the invention in any country where the patent is in force.
- Selling the invention in any country where the patent is in force.
- Using the invention in any country where the patent is in force.

A patent is a monopoly or exclusionary right that can be used to prevent others from making, using or selling the invention in a certain country. If a person does any of those things without the permission of the patent holder they can be sued for patent infringement. The selling provision is very important. If someone in the Far East makes a copy of your invention, even though you do not have a patent in the Far Eastern country, if you have patents in other countries, you can sue an importer in one of those countries on the basis of infringing the right to sell the invention.

In the case of copying, the patent owner must take a civil action to obtain damages. These actions are very expensive. Many patent owners are now taking out legal expenses insurance to cover the risk that they might lose the action and have to pay the other side's costs. There is no point in applying for patents if one does not also address the question of defence. Experience shows that most commercially successful inventions attract some form of imitation.

Patent pending

It can take some years from the time of application to the eventual grant of a patent. During this time the words 'patent pending' or 'patent applied for' are used on the product or literature to warn people that the patent applicant expects to obtain a granted patent or patents in due course. It is not possible to sue for infringement until a patent is granted.

The Workmate: showing the strength of the patent system

The Workmate is a workbench with two parts to its upper working surface. One of these can be moved in and out with respect to the other so that the two parts function as a vise.

It was invented in 1968 by Ron Hickman, a British subject. He was not an expert DIY enthusiast or a skilled engineer but saw the need for a portable workbench after he had damaged a chair in sawing.[1] His first attempt used a conventional vise and had a platform at the base that the user could stand on to improve the stability of the device. Later he made the unit portable by incorporating a folding mechanism for the base and hit on the idea of using two beams as the worktop to make a vise with jaws that moved independently of each other. A drawing, from US Patent 3,615,087 is illustrated in Figure 7.2.

He offered his invention for licence to all the major UK DIY product companies but they turned it down saying that 'it had no potential'. He then decided to put it on the market himself and sold 14 000 units mostly by mail order. At this point Black & Decker who had earlier turned it down approached him and took a licence.

Figure 7.2 US Patent for Workmate.

October 26, 1971	R.P. HICKMAN WORKBENCHES	3,615,087
Filed March 5, 1969		3 Sheets-Sheet 1

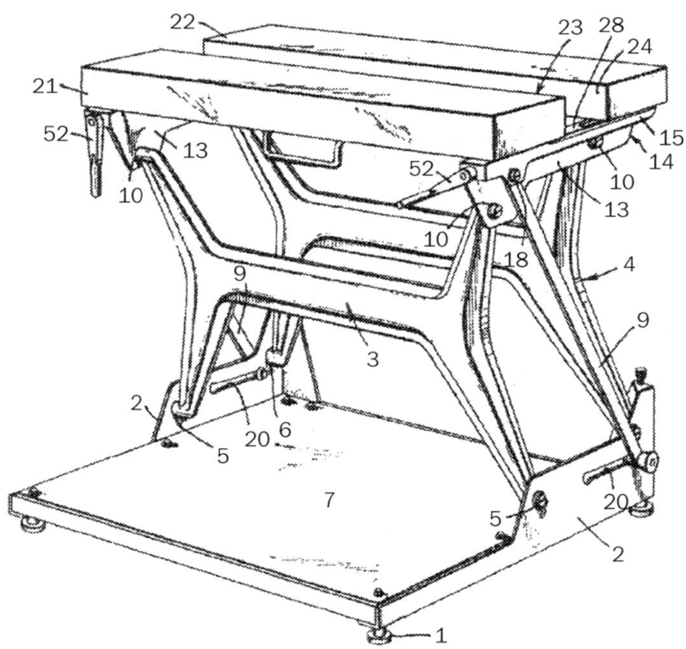

With Black & Decker marketing behind it the Workmate became an immediate success and manufacturing operations were set up in Ireland, Spain, Germany, Italy, US, Canada, Mexico, Brazil and Japan. Unfortunately, one of the inevitable prices of such success is some form of imitation. Numerous similar products were produced all of them incorporating the vise mechanism in the worktop. Patent infringement actions were successfully instituted against multiple defendants in many countries, a total of 16 cases. In one case in the UK alone, there were 10 defendants.

While this case study illustrates the strength of the patent system in protecting an invention, it should also be appreciated that without the financial strength of Black & Decker, Ron Hickman would not have been able to fight so many infringement actions.

Note

1. Roos, M.J. (1983) A case history: 'Workmate. *Les Nouvelles, Journal of the Licensing Executives Society* 18(2): 102–111.

8

Copyright, designs and trade marks

Introduction

Chapter 7 dealt with patents, the most important form of intellectual property. This chapter deals with the other types of protection for new technology: copyright, designs, trade marks and know-how (trade secrets). It also deals with the protection of software that can involve aspects of all of the above protection mechanisms and goes on to describe general IP strategies for all products.

Copyright

Copyright is about the right to reproduce one's own original work. It is claimed that the first legal judgment issued about copyright was in Ireland in AD 568.[1] The story goes that St Columcille was accused of copying an illuminated manuscript belonging to the Abbot Finian. The Abbot objected, and it is said that the case went to the High King at Tara who issued his famous judgement, 'to every cow its calf and to every book its copy'.

This established the principle that the person who produced an original work was the only person who had a right to make copies of it. Copyright is what is says, the right to copy. In modern legal terms it is called a reproduction right.

Originality

The work must be original. To qualify for copyright protection, 'original' means that the work originated with a person, and that it was not a copy of another work. The work may be literary, dramatic, musical or artistic. The definition of a literary work now includes computer programs.

In copyright a work must be original, but that does not mean unique, because copyright is not a monopoly as is a patent. This means that two people could independently conceive identical original works and each would have copyright protection. This can happen sometimes in popular music as the permutations of notes in melodies are limited. It is much less likely to arise in a complex work such as a book or a piece of software.

Authorship

The creator of a copyright work is called the author. Copyright belongs to the author, or if the author is an employee and created the work in the course of his or her employment, the copyright belongs to the employer. Copyright comes into existence automatically on creation of the work and does not have to be registered. Copyright is said to 'subsist' in a work from the moment it is written, typed or committed to some other recorded form. In copyright, the work defines itself, and it is not necessary to register a description of it, although this can be done in the USA if required. Copyright also subsists in unpublished works.

Artistic works

Artistic works are defined as paintings, sculptures, drawings, engravings and photographs, irrespective of their artistic quality. Drawings include any diagram, map, chart or plan.

Developments in technology in recent times have given rise to new methods of reproduction and to broader definitions of works that are entitled to copyright. The definition of artistic works has been extended to items such as TV programme guides. It is now considered that there is copyright in a compilation of information if significant effort went into producing it and if the work is original. This would exclude music compilation CDs, but there would still be copyright in the individual tracks.

Three-dimensional protection is limited to works produced in small numbers. An example of this would be a sculpture. If such a work is produced in commercial quantities, it loses copyright protection, and must be protected by another method such as by registered design.

Copyright precautions

In a company, it is wise to arrange the assignment of copyright to the company where there is a possibility of legal action at some time in the future which could involve the company's drawings or software. This can be done by having a clause in the contract of employment. Without this, there may be problems in enforcing the company's rights in court. Drawings should be dated and witnessed and show the copyright symbol © or the word 'copyright'. The normal format is '© Author or owner, Year'. The purpose of this is to show that copyright is claimed and thus to act as a warning to would-be copiers. However, failure to use a copyright notice does not prevent the enforcement of copyright.

Infringement of copyright

In cases of alleged infringement the onus is on the defendants to prove that they did not copy the work in question. Having dated and witnessed copies of original documents in court is clearly an advantage. The party with the earliest dated documents is clearly in a strong position. A person witnessing documents should satisfy themselves as to the identity of the author and the date of creation of the work. Precautions should be taken to ensure that original copyright documents are identified and kept safely.

In the case of software, it can be useful to print out a listing of code and get it signed and dated by the software developers. Graphic artists and others use secure, tamper-proof envelopes that can be brought to court unopened. They also use secure deposit systems.[2] Infringement of copyright can result in severe penalties, such as the handing over of all illicit copies and exemplary damages over and above any loss to the copyright owner.

Protecting ideas

Copyright protects the form of expression of an idea, but not the idea itself. Thus it may be impossible to protect a business idea. If a document is written that lays out the principles of the business idea, copyright can only prevent the copying of the document, but not of the business idea.

Reproduction rights

Copyright confers on the owner the rights to carry out 'restricted acts'. These are the right to copy the work, to make the work available to the public and to make an adaptation of the work. These are referred to as reproduction rights. Such rights refer to the work as a whole or to substantial parts of the work. Small parts of a work may be copied under the heading of 'fair dealing'. This refers to copies for private research or study, or for criticism or review, but not for any form of commercial gain.

Musical or theatrical works can give rise to other rights such as performance rights. It is also possible to have rights to translate a book or to make a film based on it. Moral rights give an author rights over the way a work is portrayed. This is designed protect an author's reputation and continues even if the work has been assigned to another owner. Moral rights also include a right to attribution, i.e. to be recognised as the author. Electronic rights allow a work to be made available in electronic form.

Design rights

Two types of design rights exist: registered and unregistered designs. Registered designs must be filed at a patent office and are examined and then granted with corresponding costs. Unregistered designs are obtained automatically, like copyright, and do not have any costs associated with obtaining them.

Unregistered design rights

An unregistered design right provides a short-term form of copyright in three dimensions for industrially produced articles. Protection is limited to 3 years from the time of placing on the market. Copying of spare parts does not infringe design rights, subject to certain limitations. The rights belong to the designer and do not have to be registered. This form of protection only exists in the EU.

Registered designs

Registered designs are more a marketing rather than a technical form of protection and cover the styling or aesthetic features of a product but not its functional features. A registered design covers the appearance of a product or part of a product, resulting from the features of, in particular, the lines, contours, colours, shape, and texture of the product itself and of its ornamentation. It excludes methods or principles of construction, or features of shape or configuration dictated solely by the function which the article performs.

A registered design gives monopoly protection and has similar novelty features to those that apply in the patent system. Registration must be made at the patent office for each country where design registration is required. However, the EU has now instituted a Community Registered Design that covers the whole EU territory in one application that is made through the EU Trade Mark and Design Office in Alicante, Spain.

Infringement can be proved without showing that copying could have taken place as would be required in the case of copyright, i.e. innocent infringement is no defence.

Registered designs are used to protect products such as domestic appliances and light fittings, furniture and cutlery. Recent changes in design legislation have extended the scope of registered designs to cover typefaces and designs that are not necessarily associated with a particular product. This means that packaging designs may be protected. Also, certain products may now be able to qualify for copyright and registered design protection, for example a printed T-shirt.

Trade marks

A trade mark is defined as: 'Any sign capable of being represented graphically and which is capable of distinguishing goods or services of one undertaking from those of other undertakings.'

The purpose of a trade mark is to distinguish one supplier's good or services from another's. This is another way of saying that a trade mark is used to connect the promotion to the supplier. If a product such as

COKE® is advertised, the intention of the advertiser is that the customer will buy a can of Coca-Cola® and not a can of a competitor's product. Thus the trade mark ensures that promotional efforts by its owner are more effective in securing sales of its own product and the trade mark itself can acquire considerable goodwill and become a valuable asset. The Coca-Cola trade marks are reputed to be worth $23bn.

Marks can be words, logos or any distinguishing mark that does not conflict with another party's trade mark. Marks that have been registered include letters and numerals, slogans, smells, shapes, sounds and images. Trade marks should be distinctive rather than descriptive, and therefore original made-up words are better than dictionary words.

Registration

It is usually wise to register a trade mark, as this gives more immediate legal protection and assists in establishing ownership of the trade mark and in licensing. In common-law countries such as Britain and Ireland, however, it is possible to take an action for passing off, and this means that it is sometimes possible to enforce unregistered trade marks. The owner must be able to satisfy the court that it has a long-established reputation or goodwill in connection with the use of the particular trade mark.

It is important that if somebody else is allowed to use the trade mark such as, for example, under license, that the reputation of the trade mark is maintained by ensuring that the goods manufactured under the trade mark maintain the same standards of quality established by the original owner. Many products are manufactured for local markets under trade mark licences. These provide that the owner of the mark can inspect the production and take samples of the products to check quality.

There are four stages in registering a trade mark: application, examination, publication and granting. It takes about 12 months to go through the procedure. The first stage is application. The applicant decides which of the 34 classes of goods and 11 classes of services the trade mark should be registered in. It is possible to register in a number of classes, but this increases the cost. The applicant should carry out a search before application to make sure that the trade mark sought is not already registered by another company. Initial searches can be carried out on the Internet by simply typing the proposed trade mark into Google. If two applicants apply for the same mark, the first to file will normally be entitled to registration.

The second stage is examination where the trade mark office checks if the mark is registerable. The mark will not be registered if it is devoid of distinctive character, descriptive of the goods or services, or is a name of a person or place. Nor will a trade mark be registered if it is likely to deceive the public or be contrary to law or morality. Once the trade mark is accepted for registration it is published in the official journal and objectors have a period in which they can oppose the application. If there is no successful opposition the registration is granted.

The symbol ® should be used to warn people that a trade mark is registered. It is an offence to use this symbol if the trade mark is not registered. For an unregistered trade mark the symbol ™ is used. Owners of trade marks should be vigilant for unauthorised use that can seriously damage their goodwill. Searching for one's own trade mark on the Internet is a good method of checking for possible infringements.

Trade marks should also be used correctly. They should not be used as the name of a product but should be used more as an adjective than as a noun. One good trade mark can protect all of a company's products and services if used correctly. Trade marks should not be confused with trade names or business names, which give no monopoly rights. Community Trade Marks provide cover for the whole of the EU and give excellent value for money.

Virgin®: the value of trade marks

When Richard Branson started selling records by mail-order in 1970 he little thought that his empire would eventually consist of 38 businesses all using the Virgin® trade mark. By 1971 he had opened a record shop in London's Oxford St and soon had his own recoding studio and record label. With signings such as Mike Oldfield, The Sex Pistols, Phil Collins and Boy George he was assured of success.

In 1984 he launched Virgin Atlantic Airways and Virgin Cargo. In 1986 Virgin Group, comprising of the music, retail and property, and communications divisions, floated on the London Stock Exchange, and was valued at £240 million. By 1986 the value had dropped to £150 million and Branson, feeling that the stock market did not understand the value of his assets, bought the shares back at the flotation price. In 1992 he sold the Virgin Music Group to Thorn EMI for £510 million.

Branson's empire expanded to cover Virgin Trains, and many other businesses including cinemas, cola, financial services, mobile phones bridal shops and hotels. In the year 2000 alone Virgin launched six new companies all in different sectors showing how a good trade mark can carry goodwill over from one business area to another. Virgin® is now registered in 27 of the 45 classes of goods and services.

Secret know-how (trade secrets)

Know-how is secret or proprietary information on producing products, and is not registered. Its protection is provided by the fact that it is kept secret. Transfer of know-how from one organisation to another is done by secrecy or confidentiality agreements. These require the receiving party to treat the information with the same care as if they had devised it themselves, and sometimes specify security precautions that must be taken by the receiving party. Know-how can include any information on manufacturing or marketing products and services and is probably the most important form of protection and transfer of technology even though it seldom appears in public. In licensing know-how, it is important that the licensee should not contract to pay for know-how that has or may come into the public domain. Five to seven years would usually be the maximum term for a know-how licence.

Terms of cover of intellectual property

- Patent: 17–20 years.
- Copyright: life of the author +70 years (EU).
- Design rights: 6–10 years.
- Registered design: 25 years.
- Trade mark: indefinite.

Protection of software

The first problem in protecting software is how to define it. It can exist in so many different forms and at different levels in a computer system.

It is not possible to protect ideas, but it is possible to protect the form of expression of such ideas when they are expressed in some fixed form such as writing, diagrams or recordings on magnetic media. Thus it is possible to protect algorithms, flow charts and source code. Direct protection is obtained using patents, copyright or secrecy. Indirect protection can be obtained by using trade marks.

Patenting software

It is possible to patent software developments (e.g. algorithms), that have novel features or which have a hardware element. Software must contain an inventive step. In the USA there should be patentable material over and above the algorithm. In Europe the patent must give a 'technical effect'. In other words it must make it possible to achieve something technically that was not possible without the software. Patents involve disclosure, but they need not disclose code. Business method patents are possible in the USA but not possible in Europe. For example, Amazon.com obtained a US patent for its '1-click ordering' but has had difficulty obtaining a European patent. A useful exercise is to examine Amazon.com's US patent 5,960,411 in the light of the above and consider whether it deserves to be granted in Europe.

Copyright in software

The reproducible expressions of software are protectable by copyright (listing of code, graphics, printouts, etc.). Copyright can be very effective in cases of direct copying or the use of unlicensed software. It becomes somewhat more difficult to identify copying where some modification of the original has taken place. In other cases it can be difficult to prove copying if public disclosure of code is required. Source code will have copyright even if it is secret as publication is not a requirement for copyright protection.

Although copyright works can pass through many forms of expression and be transmitted from a computer file to another computer file and then perhaps recorded on CD-ROM, the copyright of the work is preserved regardless of the form in which it is held. The Copyright Designs and Patents Act, 1988 states that copyright protects works 'in writing or otherwise', thus allowing for any form of expression. An adaptation of the work is also protected, but it is permissible to decompile a program if this is needed for maintenance or for interoperability.

The Internet

The Internet has raised new questions about what constitutes copying. Just because something is available through the Internet does not mean that it is permissible to copy it. The file may still be protected by copyright in that the owner of the file is unwilling to have it copied. To download a file from the Internet, a computer must make and hold a copy of it in volatile memory. It is not considered to be an infringement of copyright to view a file on a screen provided the file is not saved in a fixed memory system such as a hard disk. This type of copy is called a transient copy and is permitted. This does not permit a person to print and distribute copies of the information in the file.

Multimedia

A multimedia product is defined as a product with an interactive facility that incorporates a combination of any of the following: text, graphics, audio or video images. This gives rise to practical problems in identifying the authors of all of the works in a multimedia work and obtaining permission for the rights to reproduce each work. In designing a multimedia work, it is advisable generate as much original material as possible and not to rely on any one particular item such as a specific song or video clip in case it turns out to be difficult to negotiate the necessary rights. It is always wise to have a substitute work in mind so that the development will not be held up. Many designers of graphic and multimedia works now use file images that are bought for a once-off payment from commercial electronic libraries. The purchase includes rights to reproduce the work without further royalty payments.

Secret know-how

Secret know-how is frequently protected by preventing access to the source code. If the end-user only has a copy of the program in object code it is much more difficult to access the structure of the program. This type of protection does not prevent copying of computer disks and the use of unlicensed software. It is more suitable for specialist products where one can have a contract between developer and user to cover confidentiality issues. In practical terms, there can be difficulties of constructing dongles or resistance from users if the program is difficult to access. This subject is referred to again in Chapters 11 and 12.

Trade marks

Trade marks are very useful if you can establish a strong market presence. A company such as Microsoft derives considerable protection from trade marks such as MICROSOFT® or WINDOWS®.

The copying of products where the name and packaging is copied so as to be identical with the original product is called counterfeiting. Such products are usually distributed through the black market but can cause serious problems to the legitimate producers of brand-name clothing, software, videos, CDs and DVDs.

Intellectual property strategies

In seeking protection for any new product or technology, one should consider all the forms of IP and even use a combination of them. The first thing to do is to examine the product or technology in question and analyse its suitability for each form of IP.

If you have a product that a competitor can purchase and take apart to identify how it works, and make their own version of it by reverse engineering, you have identified a product with no secret know-how. Such a product may be protected by the patent system if it contains an invention, and you may be able to obtain a registered design for its outside appearance or unregistered designs rights for any of its parts.

Dipetane is a fuel additive that is able to give substantial fuel economies for diesel engines and central heating systems. In laboratory tests it was not possible to find out the secret formula by which it is made, or to identify the ingredients. This is a product that has substantial secret know-how and patenting it would be a mistake because this would give away the secret formula to competitors. Remember that patents cost a lot of money to obtain, maintain and defend. If you can get the same protection by keeping the technology secret, you do not need the patent system that results in disclosure of the principle of the invention.

As well as the above strategies a firm can consider trade marks and protecting the packaging design.

Process patents can be difficult to enforce, but can have good potential for licensing. One company had a process for making timber panels. A competitor started to produce similar panels but one could not say from examining the product whether it used the patented process or not. It was necessary to get into the competitor's plant to see how the panels were being made. Entry was refused, but after much argument, the company's patent agent was allowed to inspect the process. He made his report to the inventor who immediately saw that the process disclosed to the patent agent could not be economic and concluded that the process had been modified just for the inspection and had presumably been changed back the next day.

It is possible to go to court for permission to inspect a competitor's premises. If a judge is satisfied that infringement may be taking place, an order called an Anton Piller order can be given to the plaintiff who can then turn up unannounced and inspect the premises, products or process, or look for unlicensed software as the case may require.

In many cases, companies hesitate to patent inventions of limited inventiveness or market potential, fearing that the investment will not be worthwhile. Patents must pay for themselves by giving a sufficient monopoly to enable the patent holder to charge a high enough price to recover patenting and development costs. If financial forecasts do not show extra returns from patent protection, the company may decide not to apply for a patent. However, patents, even pending patents, provide a substantial deterrent to would-be copiers. They can delay attempts to design competing products and processes by as much as 12 months. It is often better to apply for patents and then drop them than not to apply.

Instructing a patent agent

Patent agents deal with the preparation and filing of applications for patents and design registrations. Many of them are also trade mark agents or have colleagues in the same firm who handle trade marks. Patent agents provide a worldwide service through their networks of colleagues in other countries so it is not necessary to seek a patent agent in each country where you want protection. They can also give advice on copyright and on licensing issues.

In filing patent applications, the job of a patent agent can be made much easier and patent drafting fees reduced by preparing a good presentation on the invention before approaching a patent agent. This document should describe the invention accurately. If it is an engineering product, it is good to provide A4 line drawings without numbers. If it is a chemical product, you should describe at least three ways of carrying out the chemical process.

It is advisable to do some searches through prior art and give this to the patent agent with arguments explaining why your invention is different or better. Identify novel and inventive features in the invention and try to show a number of different versions of it. The more ways you can show, the broader may be the patent claims. Some inventors may even ask their colleagues to come up with alternative versions.

When the patent agent submits a draft specification for approval, it is wise to examine it carefully, remembering its structure. Some parts are deliberately drafted in broad terms and some are very specific. Discuss the wording of the claims and try to see if you can get around them. It is important to identify the main commercial aspects of the invention and direct the main claims towards these features or benefits.

Notes

1. See *http://www.copyrightprotection.com/history.htm* (accessed 16/03/06).
2. See *http://www.copyrightprotection.com* (accessed 16/03/06).

9

Assessing competitors' intellectual property

Introduction

We have seen how the subject of technology acquisition embraces not just R&D but also looking at other sources of technology and products such as licensing, market-led innovation and analysis of the activities of competitors. We know that we can search for licences in the patent literature but if we find something of interest we do not know if it is available for licence or how well it is protected. We also know that our competitors are building up their intellectual assets by protecting their technology. What we need is to acquire is some competence in assessing patent strength and coverage and this subject is addressed in this chapter.

With an understanding of intellectual property (IP) from the previous chapters it is possible to assess the strength of patent claims, the duration and geographical extent of patent coverage and even check whether patents are being maintained in force by the payment of renewal fees. You can also glean large amounts of useful technical know-how that is unwittingly given away free by inventors and patent attorneys describing the background to their patent applications. It is also possible to conduct searches for expired patents and take advantage of information that is now freely available for use. We will start with a general treatment of patent searching but it should always be borne in mind that legal advice must be sought from a patent attorney before taking any major decisions related to the IP belonging to other parties.

Patent searching

Chapter 3 introduced the topic of information searches. In this chapter we apply this approach to searching through patent literature. A recent survey by PetrashWilliamson, IP management consultants found that 50 per cent of companies do not do any patent searching before starting an R&D programme.[1] Patent information is of paramount importance to people developing new products, and no R&D programme should be undertaken without conducting a patent search. We start by describing on-line searching as applied to patent databases in general followed by an example using Espacenet, which is available free through the Internet. Espacenet is the patent database of the European Patent Office.

Patents: the earliest source of information

Patent information is a very important source of technical information in general. One million patent applications are filed each year, and they must include a full disclosure of the invention in order to obtain a valid patent. Often patent literature is the only source of information concerning a particular development, as companies for commercial reasons may wish to restrict the publication of technically or commercially sensitive information until they are ready to launch their product. Also, research has shown that patent publications are often the earliest source of information, pre-dating other technical and commercial literature by a number of years. Thus, patent searches on competitors can reveal much more than looking at their websites or sales literature.

Types of patent search

There are a number of types of patent search. These are:

- 'state of the art' to see what is recent or new
- novelty searches to see what is not new
- patent family searching to see what patent coverage exists
- searches on competitors to see what they are trying to develop and protect.

A 'state of the art' search is the type of search that is carried out at the beginning of an R&D programme. The purpose of this is to give a broad technical overview of the leading developments in a particular area of technology. The search will show what areas are closed off due to having been previously patented, and will also help in avoiding carrying out an unnecessary development programme, i.e. 're-inventing the wheel'. As well as the negative aspects, however, the technical information obtained from a patent search can be extremely valuable in showing a wide range of technical options that have been tried out before, and may provide a number of technical shortcuts.

It must be appreciated that patent specifications contain information that can often be used without infringing the patent. In order to infringe the patent a company must carry out operations that are within the scope of the patent claims and the invention must be made, used or sold in patented territories during the period that the patent is in force in those territories. Many patent specifications contain large amounts of background information that are put in by patent agents to bolster the applicant's case for a patent but which do not add much to the essential features of the invention disclosure.

Novelty searches are carried out in order to assess the chances of obtaining patent protection for a particular invention. If you have invented something and there is already an invention that is very similar to it, it means that your invention is not novel. While a novelty search can be carried out on a computer database, it requires a high degree of confidence that all relevant specifications have been found. It only needs one relevant specification similar to the invention to shoot down the chance of obtaining a patent. However, if, due to inaccuracy of classification of patents, or insufficient patent searching skills, that particular specification is not found, the novelty search will give the wrong conclusion. It is useful to remember the phrase used by patent searchers that 'absence of evidence is not evidence of absence'.

In general, it must be considered that an on-line patent search is at best 90 per cent effective in finding all relevant patents in a particular subject area. Where higher levels of confidence are required such as in novelty searching, searching agencies are employed who carry out the searches using skilled searchers who are attached to the libraries of certain patent offices around the world, e.g. the British Patent Office, Swedish Patent Office and the European Patent Office. In general, a patent search using a patent office will cost between five and ten times the price of a computer patent search.

The on-line system is very effective in finding out information about specific patents of named patentees such as competitors. Equivalence searching is used to find out further information relating to a particular patent, which has already been identified. For example if you have details of the patent that has been applied for in one country, and you want to find out if it had been applied for in other countries, it is possible to find the 'patent family' quite easily from the database. It is also possible to put in the names of certain companies and find out what patents they hold.

Patent searching services are provided by a number of specialist agencies such as patent agents, technical consultants, and some universities. Patent offices, as mentioned above, are usually used for searching for novelty purposes and are more expensive.

Many parameters can be used in carrying out an on-line patent search but the main ones are as follows:

- Patent number.
- Title
- Abstract (brief description of invention).
- Patent classification.
- Patent assignee (company which owns the patent).
- Inventor.
- Publication date.
- Citations (references cited against the patent).

Example of carrying out a patent search

The following example shows how to go about carrying out a patent search. The example given is in the area of antitheft devices for cars. This example is used as it is such a simple product and one with which most people are familiar. The first type of search described here is a 'state of the art' search. There are two methods that you can use to start. The simplest is to put in key words such as 'theft' and see what you find. A more sophisticated method is to use the indexing systems.

This example is carried out using Espacenet, the patent database of the European Patent Office. This can be accessed at http://ep.espacenet.com This provides for a quick search using keywords. Here the word 'theft' was used and this produced 22 060 hits. From this it is clear that a more sophisticated approach is needed to get the number of hits down to manageable proportions while still obtaining relevant matches.

The database displays the first 10 or so of the 22 060 hits as a short description giving the title, applicant, publication number and patentee as well as the international patent classification (IPC). From this it is seen that most of the patents are classified as B60R25. As we shall see below, B covers transportation, B60 is vehicles and B60R25 is prevention of theft from vehicles.

The international patent classification devised by WIPO, the World Intellectual Property Organisation is broken down into eight broad headings as follows:

A. Human necessities.

B. Performing operations, transporting.

C. Chemistry, metallurgy.

D. Textiles, paper.

E. Fixed constructions.

F Mechanical engineering, lighting, heating, weapons, blasting.

G. Physics.

H. Electricity.

These classifications and their breakdown into more detailed areas can be studied using the classification search part of the database. If you look up vehicles you will find that the classification is B60 and if you look further under vehicles, you find 'prevention of theft', B60R25. Investigating further, you will find you will that it breaks down into subclasses as follows:

- 25/02. operating on steering mechanism.
- 25/04. preventing use of the engine.
- 25/06. operating on transmission.
- 25/08. operating on brake.
- 25/10. actuating a signalling device.

An antitheft device could be classified under two of the above headings. For example, a device that immobilises the engine could also activate an alarm. In starting a patent search you should not be too concerned about finding the exact classification initially. Thus selecting the code B60R25 would yield details of all patents in the database referring to antitheft devices for cars. This will still give a large number of hits but when combined with words in the title and abstract field it can be greatly narrowed down.

Example: patent search for 'Krooklock'

In fact starting with B60R25 yielded 38 207 hits. If you were searching for a patent for a 'Krooklock' type device that hooks between the steering wheel and one of the pedals, you could put the word 'hook' in the title or abstract field. This yielded 247 hits. You could look at these individually but it is possible to narrow the search further, for example by putting the words 'hook AND pedal' in the title or abstract. This yielded 28 hits and it would be worthwhile looking at these individually.

In carrying out a 'state of the art' search you may only be concerned with patents that have been issued in the last few years. As the patents are usually listed in order of publication with the most recent at the top of the list, it is possible to look at the most recent of the patents in any of the broader searches described above. For example, while the search yielded 38 207 patents in class B60R25, it will allow you to access the most recent 500 so you might decide to go through the titles, occasionally selecting relevant specifications for study noting their dates of publication.

If you had invented something like the Krooklock and wanted to find out if it had been invented before, i.e. to do a novelty search, you could start with a narrower category such as B60R25/02 (operating on steering mechanism) and putting the words 'hook AND pedal' in the title or abstract, eventually ending up with 19 matches. This is a manageable number to examine individually and clicking on the title will open up a page giving the abstract and a thumbnail diagram as well as other details. In the above case, one of the 19 specifications was British Patent 1,039,980 published in 1966. This is the original Krooklock patent in the name of the inventor, Kitty Zaidener and the Espacenet patent summary is shown in Figure 9.1.

Another approach is to carry out a classification search. Here you put in keywords and the search engine searches through the database and reports the most common European classifications for the description you have entered. The European classification (ECLA) is an expanded version of the International classification providing further subclasses and allowing for more detailed searches. Also, a new version of the IPC has been introduced called IPC8. This gives subclasses similar to the ECLA.

Figure 9.1 Kitty Zaidener's Krooklock patent summary

Anti-theft device for road vehicles

Patent number: GB1039980
Publication date: 1966-08-24
Inventor:
Applicant: KITTY ZAIDENER
Classification:
– international: ***B60R25/02***; **B60R25/02;**
– european: B60R25/02C2
Application number: GB19620048692 19621227
Priority number(s): GB19620048692 19621227; FR19630958556 19631226

Abstract of **GB1039980**
1,039,980. Preventing theft of vehicle. K. ZAIDENER. Dec. 20, 1963 [Dec. 27, 1962; March 7, 1963], Nos. 48692/62 and 9188/63. Addition to 1,007,711. Heading B7J. A theft-preventing device for a motor vehicle comprises a telescopic member 10 of square, elliptical, or round section, comprising a tube 11 having a hook 12 at one end and a rod 13 sliding within the tube and being provided with a hooked end 14 protected by a rubber sheath 17 and having a series of indentations 16 which may be engaged by a spring-loaded ball to locate the rod 13 at any length of extension. A lock 18 is riveted to a plate 19 fixed to the tube 11 and comprises a cam disc 22 which may be rotated to press a ball 24 into engagement with one of the indentations 16 whereby the device is locked at a given extended length. A spring-loaded ball 25 is also located in the lock casing for the initial location of the members 11 and 13. In use, the hook 12 is placed about a control pedal and the hook 14 about the steering wheel. The lock may be fixed in other positions and may be a spring-loaded push lock releasable by a key. In a modification, the hooks may be replaced by "hinged handcuff-type members".

1,039,980.

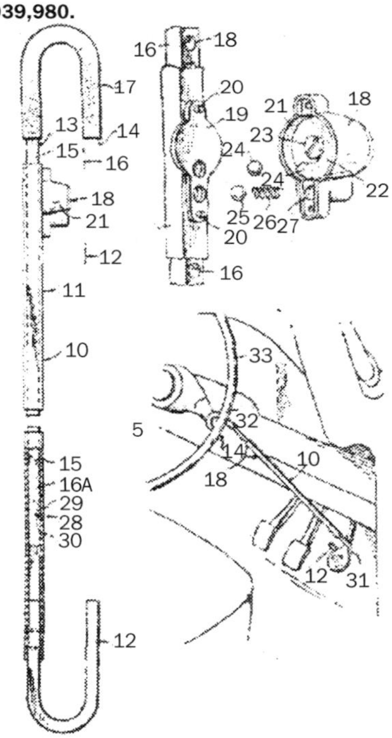

In patent searching a little knowledge is a dangerous thing and it is worthwhile to avail of the excellent training programmes and materials available from the EPO. You can attend a seminar at the EPO office in Vienna, or locally or learn via the Internet at http://patentinfo.european-patent-office.org/learning/

You can carry out more sophisticated searching on the subscription databases. You can search the full text of patents and not just the title or abstract. You can look for combinations of keywords or keywords in context. This type of searching can also be applied to journal literature and you can search a number of databases, both patent and literature, simultaneously. A list of databases for patent searching is provided in Appendix 1.

Copies of actual patent documents can be downloaded, printed out or ordered from some of the databases or through technical or patent office libraries.

Searching for equivalents or patent families

Most of the patent databases provide cross-references to the family of patent documents that emerges as an invention is examined in different countries. These documents follow a certain format agreed with WIPO. The country code comes first, followed by the number of the application or granted patent. Finally, there is usually an 'A' or 'B' version of the specification, A referring to an application and B referring to the equivalent granted specification. A2 or A3 can refer to later publications of the application, the later ones containing the search report.

The evolution of a patent family is shown in Figure 9.2 where different specifications emerge depending on the filing route chosen. It can be seen that an application filed in the UK as a priority document gets a 'GB' prefix and then can proceed down one of three routes. If it goes straight into a national patent office such as the USA, it will emerge as an application and then as a grant. If it goes into the European Patent Office it will get an EP prefix and if it goes into the PCT system it will be published first with a PCT prefix and later as a WO, the WO referring to WIPO and not the world, as there is no such thing as a world patent and the PCT being a filing system only.

Figure 9.2 also illustrates some other country codes, JP being Japan and AU being Australia. Most codes are obvious such as FR, NL and DE but some of the less obvious ones are AT for Austria, CH for Switzerland, CN for China, and ES for Spain.

Assessing the strength of a patent

As explained in Chapter 7, the scope of the monopoly obtained is determined by the claims of a patent. These carefully drafted sentences are structured in a defensive layer, the first claim being the strongest and

Figure 9.2 Evolution of a patent family

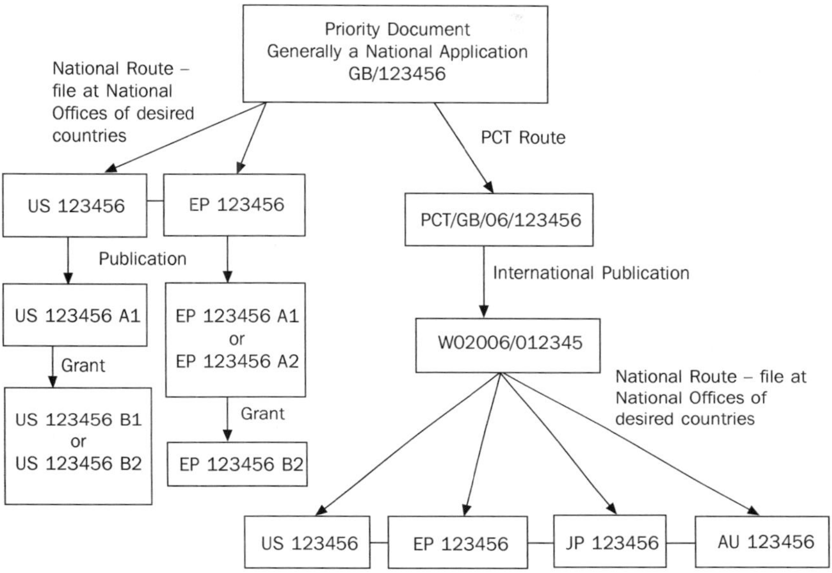

therefore the most important. The other claims usually try to define the invention in different terms or focus on other key features.

The following is the wording of a patent claim:

> I claim: a vibration sensing device comprising: a non-conductive base support member; a pair of spaced apart electrically conductive plates upstanding from the base support member and electrically insulated from each other; an annular track formed in each plate by a hole symmetrical about its central axis; and a single elongated electrically conductive bar, symmetrical in cross-section mounted between the plates on the annular tracks, the cross-sectional area of the bar relative to the area of the hole being such as to permit movement of the bar off the annular tracks on vibration of the device, the said bar and annular tracks forming part of an electrical circuit.

A useful exercise is to try to draw this device. Do this without looking at Figure 9.3. This gives an insight into the drafting process and shows how important it is to choose the correct words and the features to protect. What do you think it might be used for? If you wish to study this patent further you should search for US Patent 4,185,180 on one of the databases. This would be useful when reading the next paragraph.

Figure 9.3 US patent 4,185,180 front page

United States Patent [19] [11] **4,185,180**
Anderson [45] **Jan, 22, 1980**

[54] *VIBRATION SENSING DEVICE*
[75] Inventor: **Francis J. Anderson, Dublin, Ireland**
[73] Assignee: **Institute for Industrial Research & Standards, Dublin, Ireland**
[21] Appl. No. **919,519**
[22] Filed: **Jun. 27, 1978**
[30] **Foreign Application Priority Data**

Jun 27, 1977	[IE]	Ireland	1306/77
Jul. 29, 1977	[IE]	Ireland	1584/77
Oct. 12, 1977	[IE]	Ireland	2086/77

[51] **Int. Cl.^{2} H01H 35/14; G08B 13/00**
[52] **U.S. Cl. .... 200/61.45 R; 200/61.52** 200/61.93; 200/277
[58] **Field of Search**...... 200/61.45 R, 61.48-61.53, 200/277, 61.93, DIG 45; 340/566, 683, 541 545, 547, 565, 669, 670; 102/262, 272; 73/570 632, 649, 651, 652

[56] **References Cited**

U.S. PATENT DOCUMENTS

2,721,611	10/1955	Gordon	200/61.52X
2,996,586	8/1961	Shindledecker et al..	200/61.49 X
3,161,737	12/1964	Hall	200/61.52
3,161,738	12/1964	Hall	200/61.52
3,295,101	12/1966	Ellis	200/61.45 R X
3,486,143	12/1969	Wilkes	200/DIG. 45
3,696,380	10/1972	Murphy	200/61.93
4,025,744	5/1977	Kniskern	200/61.45 R
4,085,304	4/1978	Hasler	200/61.45 R

FOREIGN PATENT DOCUMENTS

2423660	12/1974	Fed. Rep. of Germany	200/277
776878	11/1934	France	200/61.45 R

OTHER PUBLICATIONS

Wilkes, D.F.; "Rolamite: A New Mechanical Design Concept"; Sandia Laboratories, Mar. 1969; No. SC-R-R-67-656B, pp. 59–61.

Primary Examiner – James R. Scott
Attorney, Agent, or Firm – Ladas, Parry, Von Gehr, Goldsmith & Deschamps

[57] *ABSTRACT*

The invention provides a vibration sensing device which is used in safety and security equipment. The device is effectively an electrical switch and includes a pair of spaced apart electrically conductive plates each having an annular track formed by a hole. An electrically conductive bar is mounted between the plates on the tracks. On sensing a vibration the bar will resonate lifting off the tracks thus making and breaking the electrical circuit between the plates.

16 Claims, 25 Drawing Figures

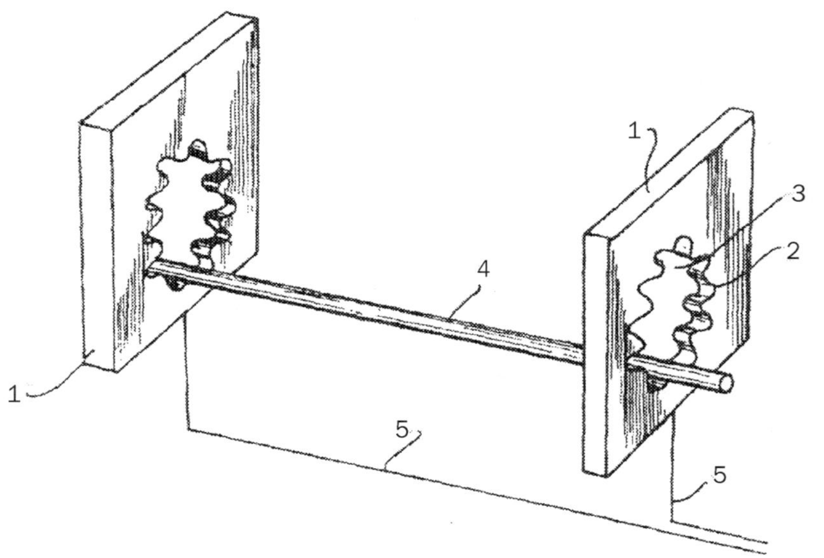

Reading a patent specification

Read US Patent 4,185,180 or look at Figure 9.3. The front page of the specification gives the filing, indexing and searching details together with an abstract of the invention and where possible a representative drawing. The numbers before the different headings are used internationally. For example, the priority date is always number [30]. So if you come across a patent in a foreign language you will know that [30] refers to the priority date.

Number [51] refers to the international classification. Here you can see that it is classified as H01H 35/14 acceleration switches, and G08B 13/00 intruder alarms. The invention is a very successful product for detecting forcible entry to a building using the fact that the acceleration caused by the impact of the intruder on a door or window will break the contact between the bar and plates. Other items of interest on the front page are the references cited that cover both US and foreign patent documents and 'other publications' outside the patent literature entirely.

The text of the patent has a particular structure that is not obvious even to a technical expert. The remainder of the document is laid out as follows:

1. Preamble: background, problem to be solved (column 1) and prior art (columns 2–3).
2. Statements of invention: legal statements of what the invention is, very broad with objects, i.e. advantages (column 3).
3. Description of drawings (column 4).
4. Specific embodiments: detailed descriptions of how the invention works, very narrow and specific (columns 4–7).
5. Broadening-out section: attempts to protect other versions and wider applications.
6. Claims: legal statements based on the statements of invention, very broad (columns 8–10).

The term 'prior art' refers to previous inventions or publications of relevance. Here the patent agent is alerting the patent examiner to the existence of certain inventions and pre-empting any objections by explaining why they are different or do not achieve the same results.

The term 'embodiments' is used by patent agents to describe versions of the invention. The first example described is usually called the 'preferred embodiment'. This description is always quite narrow and specific. It can be seen how the patent agent has tried to broaden the scope of the invention by showing other embodiments in figures 4–10.

It is important to recognise that some parts of the specification are drafted broadly so as to get maximum protection, while others are quite narrow as they need to describe accurately how the invention works.

Types of patent claims

There are two types of patent claims, dependent and independent claims. The dependent claims always refer to a previous claim so will start with the words 'I claim ... as in claim 1' or as in another claim. In other words these claims are subsets of previous claims and are really a kind of shorthand that avoids writing out the whole detail of the earlier claim each time. Dependent claims being subordinate to independent claims can be regarded as having less legal scope. Independent claims are usually the first claim but there may be more claims further down the list where the patent attorney has attempted to describe the invention in a completely different manner or from a new starting point with no reference to a previous claim. All of these independent claims must be considered first when trying to assess the strength of a patent.

If you see the words 'characterised in that' or 'wherein' in the middle of a claim it means that what has preceded these words is recognised as prior art and the invention refers to the features following those words. This usually refers to inventions that are developments of, or improvements to previous inventions.

Strength of claims

An initial assessment of the strength of the claims in a patent is obtained by careful study and interpretation of all the independent claims of a patent. From this one attempts to see if the claim appears to close off all possible technical solutions.

For example, if we go back to claim 1 of US 4,185,180, we could test it by asking if there is any other way that such a device could work without infringing the patent. This is done by trying to see if any of the principal features of the claim could be modified or omitted while still giving a workable solution. In this case a competitor decided that the 'pair of spaced apart electrically conductive plates' and the 'annular track' could be replaced by gold plated wires to give a workable device that did not come within the scope of the claims.[2]

It would be unwise to assume that all patents are open to circumvention. Most are not. However, the above case illustrates the point that inventors sometimes see only the one technical solution. Patents of broader scope can be obtained if the inventor calls in colleagues and asks them to try to find other technical solutions. If this can be done, a patent agent may be able to draft broader claims covering the principle of the invention rather than a specific version of it.

Using patent search information

Patent searches reveal both application and granted patent documents. When application documents are published, they can reveal a lot about a company's R&D plans and give details of products that may not yet be launched. By looking for search reports it is possible to see what patents have been cited against the application. Examining these can often show that the patent is unlikely to be granted with the broad claims initially filed.

The list of citations appears on the front of US and European patents, as in Figure 9.3, but at the back of European patent applications (A2 or A3 documents).

By looking up the references cited, it sometimes becomes clear that the patent is not very strong and that there have been previous patents covering earlier versions of the same technology. The patent in question may be just a minor improvement on the earlier patent. In many cases it is found that the earlier patent has expired or is no longer in force because renewal fees were not paid. This means that it is public domain and can be used with impunity provided the improvements are avoided. It is possible to check the legal status of a patent by using the INPADOC database that is accessible through Espacenet.

Looking at patent families shows the extent of territorial coverage of each invention. It was pointed out earlier that many patents have limited territorial coverage and thus the invention can be exploited provided it is not made, used or sold in territories where patents are in force. The age of a patent should also be examined. Many patents are no longer in force due to having reached their expiry date. Some firms carry out searches for expired patents by searching for patents that are more than 20 years old. This can be done by constructing your search specification to set the publication dates so that you find all patents published before a certain date. This is particularly common in the pharmaceutical industry where many drugs still have a market long after patents have expired. These are called generic drugs.

Legal advice

There is much that can be done by a technology manager before it becomes necessary to seek the advice of a patent or legal professional. In this and previous chapters it can be seen that IP is subject where a manager can develop basic skills and do a considerable amount of work, mainly in the form of desk research by taking advantage of the information services that are now becoming available. It is possible to obtain basic information about competitors concerning the scope and strength of their IP and to investigate the prospects for obtaining protection for inventions, either your company's or those revealed in competitors' patent applications. However, it is very important to know when to call in the experts. When you have reached a decision point concerning going ahead with a project that relates to the IP of another party it is essential to obtain legal advice. This can include advice on the correct interpretation of patent claims or confirmation of the extent of patent coverage, or the validity of a competitor's patents in the light of prior art.

Patent litigation is notoriously expensive and no firm should allow itself to run the risk of being drawn into a lawsuit. Threats are frequently made by competitors that turn out to be groundless but it is important to know when you are on safe ground. Patent law provides remedies if a competitor makes a treat that is groundless and causes you to cease your activities. A technology manager does not need to be a lawyer but a little knowledge goes a long way in preventing expensive mistakes. The most important thing to learn is when to obtain advice.

The ethics of copying

Many companies now have stated ethical policies that can be seen on their websites. These cover relations with their employees, the local community, customers, suppliers and competitors. Many state that they will seek to compete fairly and will not deliberately seek to damage a competitor. So where should a company stand on issues concerning the

IP of a competitor? This is an evolving area and the following are suggested guidelines that a company could consider.

Suggested guidelines on copying

A company should not deliberately infringe the IP rights of a competitor. This means that a competitor should not be targeted or their products or technology copied where these are protected by IP rights.

Where a competitor has not sought protection such as by failing to obtain patents then it is legitimate to copy provided no other rights are infringed. Thus if a competitor has not obtained patents in certain countries, it is legitimate to copy their products and technology in those countries.

Where patent protection is limited because possibilities exist to design around the claims of a patent, it is legitimate to do so.

While many activities may be legitimate, a company should always consider if they are also fair. For example, a large company should not use its financial and legal strength to target a weaker competitor by, for example, trying to prove that their patents are invalid, or by threatening them with an infringement action where there is some doubt about the merits of the case.

Notes

1. See *http://www.ipambestpractices.com/index.html*
2. It should be noted that US 4,185,180 has now expired and so knowing how it was circumvented is only of interest to illustrate the point.

10

Building intellectual assets

Introduction

In earlier times wealth was mainly created by manufacturing products. Nowadays, wealth is often created from technology, from the application of science but not necessarily through manufacturing. Companies can make substantial sums by licensing their technology, and information technology can create new industries with products such as software that can be sold over the Internet without ever appearing in concrete form. The product of the brain is becoming more important than the product of the hand. This chapter deals with intellectual assets, the intangible assets of companies, and provides a framework and techniques for managing these assets.

Traditionally, businesses were valued on the basis of the assets shown on the balance sheet. These assets were fixed assets such as buildings and machinery and current assets such as inventories, cash and debtors. As technology has advanced, the asset value of a company as shown by its balance sheet and its real market value have diverged, showing that wealth is being created from something that is not being adequately accounted for in looking at the tangible assets of the business.

An analysis of company acquisitions in the USA shows that the average company was acquired for 4.4 times its book value, and that for technology or knowledge-intensive companies the ratio of market value to book value was over ten times. The average over all industries was six times. In other words, only 15 per cent of a company's value is being captured by current accounting methods. This difference is accounted for by the value of a company's intellectual assets, its intellectual capital.[1]

Definition of intellectual capital

Intellectual capital is defined as the difference between market value and book value.

$$IC = MV - BV$$

Edvinsson and Malone in their book *Intellectual Capital: The Proven Way to Establish your Company's Real Value by Measuring its Brainpower*[2] state that intellectual capital has two main components: human capital and structural capital. Human capital is: 'The combined knowledge, skill, innovativeness, and ability of the company's individual employees to meet the task at hand. It also includes the company's values, culture, and philosophy. Human capital cannot be owned by the company.'

Structural capital is defined as: 'The hardware, software, databases, organisational structure, patents, trade marks, and everything else of organisational capability that supports those employees' productivity – in a word, everything left at the office when the employees go home. Structural capital also includes customer capital, the relationships developed with key customers. Unlike human capital, structural capital can be owned and thereby traded.'

The hidden roots of value

Edvinsson and Malone describe intellectual capital as 'The hidden roots of value'. Like a tree, only the trunk and branches are showing above the ground. The root system, which nourishes the tree but cannot be seen, represents the hidden assets.

The concept of intellectual capital has relevance in all sectors of industry and business and not just the high tech or IT sector. The principles were pioneered by Edvinsson in his own company, Skandia, the biggest financial services company in Scandinavia. He was appointed as director of intellectual capital in 1991 and produces an annual report showing the value of its intellectual assets.

Another company that has pioneered this approach is Dow Chemical, which appointed Gordon Petrash as director of intellectual asset management (IAM) in 1993. Dow Chemical define intellectual capital as 'knowledge with potential for value', while intellectual assets are defined as codified knowledge or knowledge that has been written down and is shared across the business. These definitions recognise the increasing importance of knowledge, and most business are now knowledge based. The distinction between intellectual capital and intellectual assets is that

Figure 10.1 **Intellectual capital: the hidden roots of value**

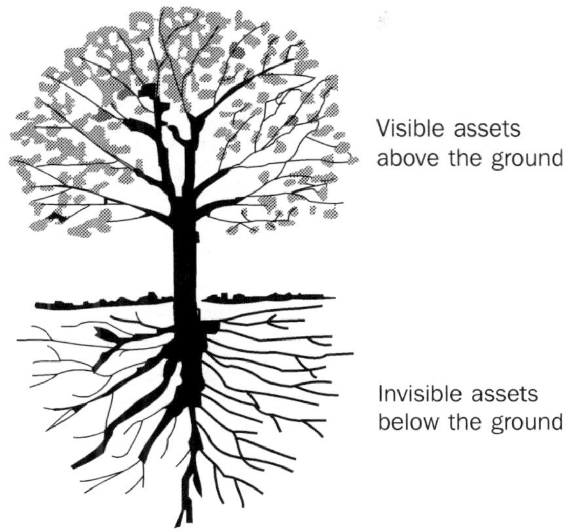

intellectual capital looks to the future, while intellectual assets have already been created. Dow define intellectual capital management as 'the process of proactively managing, protecting, leveraging and reporting all knowledge assets to better enable us to gain/sustain competitive advantage and maximize future value.' Another way of looking at intellectual capital is to describe it as the other side of the balance sheet that matches intellectual assets.

Intellectual capital management

Intellectual capital management is seen as being more strategic and long term. It recognises the value of intellectual assets and seeks to maximise the long-term return on these assets. It also seeks to value them and the business by this yardstick. Many problems arise in accounting methods of valuation, however. Traditionally, accountants recognised the gap between book value and market value as being made up by intangible assets or goodwill. Often this only appeared in the balance sheet if some external valuation criteria were applied, such as the sale of the business. The tendency to continue to value businesses very conservatively still exists and can produce serious distortions. For most companies, their annual report says nothing about the real value of their intangibles and makes no attempt to bridge the gap between book value and market value.

In many companies, research and development is charged as an expense and serves to reduce the profit reported. In more progressive companies it is capitalised or treated as an investment and amortised or written off over a number of years when it is expected to provide a return. In some cases accounting rules prevent this. When IBM bought Lotus for $3.5bn in 1995, almost $2bn of that amount was paid for technology under development. Yet IBM was required to charge the $2bn as an expense, which reduced its reported profits and could not report it as an asset on its balance sheet.

There are signs that the picture is changing. The Financial Accounting Standards Board in the US is at last recognising that more realistic valuations of intangible assets must be allowed. Markets react instantly to announcements by major technology companies of new technical developments. Suddenly the potential of an intellectual asset starts to be revealed. Some new methods of valuation will have to be developed by the accounting profession to keep pace.

Intellectual asset management

IAM is a subset of intellectual capital management. It is more focused on the intellectual assets that already exist and provides a set of techniques for maximising the return on intellectual assets. Gordon Petrash, who had built up the IAM side of Dow Chemicals, moved on to PricewaterhouseCoopers where he built up a consulting arm in IAM. He was then recruited by Delphion to develop this side of their business. Delphion had purchased the IBM patent database and had extended it to cover other technical information sources by linking up with the Thomson Corporation, another information provider that owns Dialog. Here we see concrete signs of the knowledge economy, the building of links between technology, information services and the management of knowledge. Petrash went on to set up PetrashWilliamson providing information and services on IAM.[3]

IAM uses aspects of knowledge management, which is defined as 'the process of getting knowledge from the people who have it to the people who need it, at the right time, to satisfy a business need.' It relies heavily on information systems and in particular on having access to databases of information about technology. It provides management techniques for keeping track of intellectual property (IP), and for its exploitation.

Forms of intellectual assets

Intellectual assets are more than just IP, which is the part of intellectual assets that has been captured in some form. Intellectual assets also cover the information, skills and experience of a company's employees. These assets can also be more dynamic, covering the way the skills are focused and utilised by good management practice, helping to develop customer and supplier relations and encouraging creativity and innovation.

The structural aspects are the systems that are set up to support the human aspects. They include the organisation structure, information systems, technology and IP. They also include the structures that have been set up with customers and suppliers, and joint ventures or development programmes with other firms.

The management of the human intellectual assets (which are not owned by the business) requires one approach, while the management of the structural assets (which are owned) requires another. The management of the human assets is about developing an innovative culture in a firm. Aspects of developing an innovative culture were dealt with under the heading of creativity in Chapter 4.

The management of structural assets covers a wide range of activities from information systems to the acquisition of technology and more. In this chapter we are focusing on intellectual assets that have been codified or written down in some communicable form. These intellectual assets are a company's intellectual property (IP).

Intellectual property management

IP is owned and can be traded. It should be treated as a resource, which, like any other resource, can be acquired, developed, maintained and exploited. These aspects will be described briefly in this chapter and will be developed in more detail in later chapters.

Acquisition of intellectual property

IP is acquired through the process of technology acquisition. If the technology is developed in-house, various forms of statutory IP such as patents, copyrights and designs will arise. The company decides on the advisability of investing in forms of protection that require registration, such as patents, designs and trade marks. Once applications for these have been filed, they must be followed up, pursued through the system and secured as granted IP protection. Copyrights are created automatically in literary and artistic works such as software and drawings. These must be identified, original documents kept safely and provisions made for keeping track of changes.

When technology is acquired externally it can be in the form of rights that are purchased such as the buying of a patent. This is treated exactly the same as if it originated from a development within the company. Provisions have to be made for assessment and transfer of the IP rights and for their use. Often, external technology is acquired by licence. Instead of being owned, it is rented. This also requires management of the IP even though it still belongs to the licensor.

Development of intellectual property

Much IP arises from development projects but IP itself can be actively developed by encouraging employees to be aware of opportunities for patenting and otherwise protecting the company's technology so as to assist in commercial exploitation. This requires training courses in IP and management systems that encourage and reward new ideas especially those that are patentable. IP managers can also develop their portfolios by acquiring strategically important patents, by making sure that they obtain wide geographical coverage and by supervising the patent prosecution process to ensure that patents of significant strength are obtained.

Maintenance of intellectual property

Patents, designs and trade marks need to be maintained in force by the payment of renewal fees. While reminders are issued by patent agents, proper IP management requires a company to be aware of the renewal dates and to have regular monitoring of IP to ensure that informed decisions can be taken on what is to be kept in force and what is to be abandoned. Many companies do not know what patents they have and some have lost their patents by neglecting to pay renewal fees in time.

A second aspect of IP maintenance is monitoring competitors to spot and stop infringement or copying. Systems need to be set up so that all employees are aware of what the company's IP actually covers and are on the lookout for any infringing activities. The company also needs a policy on IP defence. If it is serious about enforcing its rights, it needs to ensure that funds are available for this purpose. Smaller companies can use IP legal expenses insurance.

Exploitation of intellectual property

Commercial exploitation of IP includes manufacturing and licensing. In manufacturing, the IP protects the products and technologies against copying and allows a higher price to be charged when the products are sold. This increases profits and justifies the cost of the IP. As well as manufacturing, many companies also licence out their technology. This can bring in substantial royalty income, which is seen by investors as indicating significant value for the company's technology and future prospects.

In 2002 Gordon Petrash wrote 'Patent revenue is extraordinarily profitable. There is often almost no cost involved in licensing out an existing patent to generate revenue – and the same patent may be licensed many times. For example, IBM has a 90 per cent profit rate on $1.7 billion of annual patent royalty income. So while this represents only 2.1 per cent of IBM's revenue, it is 17 per cent of IBM's pre-tax profit.'[4]

Licensing out requires the setting up of a dedicated activity to evaluate the potential for licensing out, to identify prospects, to promote the sale of the technology and to manage the transfer. These points will be dealt with in more detail in later chapters.

Organisational aspects of intellectual asset management

In many large companies, patents are filed through the corporate legal department. Once the R&D department has submitted an invention disclosure, they tend to move on and leave it to the lawyers to get patents granted. R&D scientists are judged by the number of patents they hold but not by the number of patents that are generating revenues. Similarly, the legal departments regards patents as a legal matter and do not consult with commercial departments on exploitation issues.

In 1996 Procter & Gamble examined its 27 000 patents and found that they were exploiting less than 10 per cent of them and that many other leading companies were in the same situation.[5] Dow Chemical set up a review structure whereby its patents were regularly monitored to see if they were being exploited. It found that many patents were classified as 'orphans'. No department had responsibility for them. One of the key elements in the Dow IAM strategy was to find a business unit to take responsibility for the exploitation of these patents. Patents were classified as: 'using', 'will use' or 'will not use'. From this the company then decided if the patents should either be put into use, licensed-out or abandoned. In a period of 10 years Dow saved $40 million in unnecessary patent fees by this strategy.

In the McKinsey Quarterly, Meagan C. Dietz and Jeffrey J. Elton state that many companies lack the processes and organisational structures to deal with IP management and to link it to their strategic objectives. They use poor metrics to define performance often rolling licensing revenues into existing business unit activities and hiding the value of IP and de-emphasising its management. In the article they also emphasise the importance of combining licensed-in technology with doing R&D to generate extra revenues.[6]

The key to successful IAM is to have a multi-disciplinary team which is responsible for the continuing review and commercial exploitation of projects and related patents. This may be the innovation committee in smaller companies or a specialist team in larger ones. Metrics should be established for performance in IAM. These could include revenues or return on investment from R&D, patenting and licensing, number of new products launched, average age of patents, patent coverage in relation to business coverage and patent coverage in relation to competitors.

Using the patent databases it is relatively easy to collect statistics on the number of patents filed per index class. Thus it is possible to construct patent radar maps as shown in Figure 10.2. These compare number of patents filed by a company in each technology area with its competitors.

Sophisticated patent management software programs are now used by many companies and universities to keep track of their patent filing, prosecution and exploitation activities. Some of these will also produce radar maps, patent ageing schedules and citation trees showing whether a company's patent has been cited against a competitor. In such a case it may be able to force the competitor to take a licence.[7]

Valuation of intellectual assets

Intellectual assets are valued according to two basic approaches: the cost of replication approach and the potential income approach.

Cost of replication

This approach examines a technology and its IP and consists of an assessment of what it would cost to develop the technology and IP in-house. If the technology is already owned by the company, such an assessment is done by using financial records of the development and

Figure 10.2 **A patent radar map shows technology areas patented by companies. Company A has a much wider spread of technologies than company B**

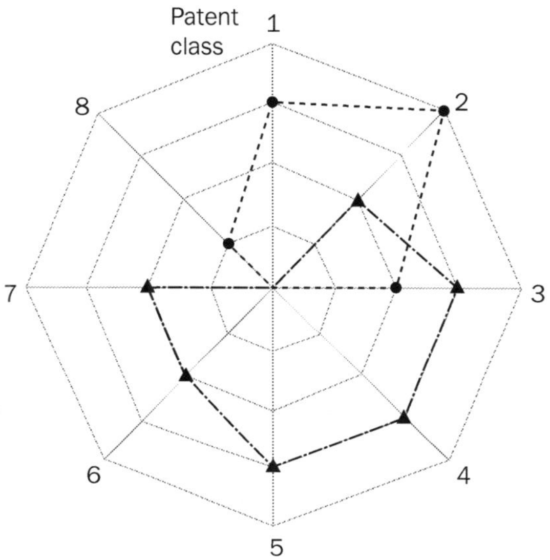

protection of the technology. If the technology is not owned, the cost of an R&D programme must be estimated and an assessment of the IP carried out to see if it is possible to design around it and to assess how much this would add to the development cost. This type of estimating is very difficult and some experimental work is often carried out to assist in determining the costs and risks.

Potential income approach

This approach assesses the potential income of the technology over a number of years and uses the present value of this income stream as an indicator of the value. Present value represents the amount of money you would have to pay today as a lump sum to obtain a future income stream over a number of years. This approach needs accurate sales forecasts for the products involved. If a company already owns the technology, such estimates will already be available as they would have formed part of the original decision to develop the technology. This information, updated and combined with price and cost estimates, enables the future profit to be calculated. If the company does not own the technology, some or all of the information will have to be estimated.

In addition to these approaches, which are quantitative, the principles of technology impact analysis are also used to give qualitative estimates of technology value by using scoring methods as discussed in Chapter 5. Quantitative and qualitative techniques would usually be used together before taking decisions on purchasing technology. Other techniques relating to evaluation of licences will be discussed in Chapters 12 and 13.

Financial analysis of technology

The cost of replication only establishes a minimum value for a technology. Such a technique is likely to undervalue the technology by not taking its potential to generate income into account. This method of valuation might be used in acquiring the technology in a fire sale where a company had run out of cash and had to sell off its technology under pressure from creditors.

The potential income approach can reveal the true value of a technology but requires a lot of assumptions about future sales and profits. Any method of financial analysis is only as good as the quality of the information put in. Unfortunately, it is hard to get good quality information if the technology is at an early stage in its life or if you are evaluating it as an outsider and are not given information known to the developers.

Financial methods usually result in a number. This may be a value in money or a payback period in years. Because of the sophisticated techniques involved in calculating the answers, people assume that these numbers are correct and treat them with great respect. However, in many financial models, by increasing sales volumes or prices by say 20 per cent, the value of the technology may increase by 50 per cent. Similarly, a reduction of 20 per cent may reduce the value by as much as 50 per cent. This extreme sensitivity of the answer to changes in the input data is a cause for concern, so many analysts use a technique called sensitivity analysis. In this technique, different calculations are made taking assumptions above and below the forecast sales and profits to see how much the price will vary.

Using sensitivity analysis, a purchaser of a technology is able to examine the impact of a reduction in sales, for example, and get an idea how much the price calculation is likely to vary. If the price swings widely for changes in input data, the purchaser will treat the figure as negotiable within a wider range.

Present value

Present value is a concept that recognises the time value of money. If you have £100 today and invest it in the bank at 6 per cent interest it will be worth £106 at the end of the year. Therefore £106 in 1 year is only worth £100 today. This is called discounting. The further into the future a particular sum is received, the less it is going to be worth today. Discounting is applied to the projected income that is expected from a business or technology, with the result that income in 5 years' time would not be given equal value to income received in 1 year's time.

Table 10.1 shows a simplified financial projection for the performance of a new technology. The technology results in the sales of a new product and the projected sales figures are shown increasing from £5m to £500m in 5 years. In each year the cost of sales deducted from the sales revenue gives a profit that can be seen to grow from zero in the start-up year to £200m in year 5.

If you did not use discounting, you would calculate the present value of the technology as the accumulated profit which, if you took 5 years, would amount to £360m. If you took a longer period, the value would be many times this, so when do you stop? In the line below, the cumulative discount factors are shown at 6 per cent. When these are applied to the profit figures it can be seen that the cumulative value after 5 years is now £279.5m, much less than £360m without discounting. This is called the present value. The discount factor at 6 per cent for 10 years is 0.558 so you can see that values this far out count much less.

Using this technique, uncertain values far into the future are given less validity, and a present value calculation can be made that does not increase significantly if you extend the projections over a long number of years. The technique is called discounted cashflow and the present value is calculated over a number of years at a certain discount rate, which is the equivalent of an interest rate but used in reverse.

Table 10.1 Simplified table of financial performance in the first 5 years of a new technology

	Year 1	Year 2	Year 3	Year 4	Year 5
Sales	5	50	150	250	500
Cost of sales	5	40	100	150	300
Profit	0	10	50	100	200
Cumulative profit	0	10	60	160	360
6% discount factor	0.943	0.890	0.840	0.792	0.747
Discounted profit	0	8.9	42.0	79.2	149.4
Cumulative discounted profit	0	8.9	50.9	130.1	279.5

The example illustrated is greatly simplified so as to explain the basic concept of discounting. In a real situation, there would be start-up costs and these would produce minus sums at the beginning before any profits are earned. These would be deducted to give what is called the net cashflow and the present value is called the net present value or NPV. For example in an R&D project there might be an investment of £50m before the project starts. This goes in as a negative cashflow item at the beginning. For this kind of project you use discounted cashflow (DCF) for investment decisions. If you invested £50m in R&D and received a return with a present value of £279.5m it is clearly a good investment.

It can be seen that the technique is useful for making investment decisions in a company. Investing in an external technology is somewhat different, because instead of an R&D cost there is an asking price for the technology. You put the asking price in at the beginning as if it was an R&D cost and see if the NPV of the return is greater than the asking price. You may also have start-up costs that occur in the initial years.

The DCF technique, though widely used by accountants, is nevertheless criticised for often giving values that turn out to be misleading when making decisions on purchasing technology. This happens not because of a defect in the technique but because the sales and profit forecasts are incorrect or because of the discount rate chosen.

Selecting a discount rate

In the example above, the discount rate is the same as the interest rate. If you choose such a rate you are effectively comparing the return on the investment in the technology with the return you would get if you put the same sum of money in a deposit account in a bank. Many companies require a minimum return on investment and may use a discount rate of say 15 per cent. Others argue that the discount rate should reflect the risk in a project and may select a discount rate of 40 per cent. The best approach is to take a discount rate that reflects the company's cost of capital.

One method avoids choosing a discount rate by doing the calculations in reverse. This is called the Internal Rate of Return (IRR) method. This method asks what is the rate of return at which the NPV is zero? The answer is a percentage that can be useful for comparing competing projects but is of limited value in purchasing technology externally.

Accuracy of forecasts

No forecast is ever exactly right. Some are accurate and some turn out to be very wide of the mark. To understand how these discrepancies might arise, we must decide what we mean by a forecast. In many cases the sales forecast is the best estimate calculated after careful analysis by the marketing experts. Does this include a probability factor for commercial success? Usually not. It does not consider what the sales will be if a competitor enters the market, or if people have different expectations for the product that are not fulfilled. We must also ask if the figures include a probability for technical success. What happens if the product turns out to be unreliable, or if there are development delays or production difficulties that mean that sales revenue is late in starting?

Expected commercial value

A technique that calculates the expected commercial value attempts to deal with technical and commercial uncertainties by applying a probability of technical success factor to the technical aspects and a probability of commercial success factor to the commercial activities. This technique has the advantage of separating risk estimates from forecasts and discount rates. You can take the sales and profit forecasts and a realistic discount rate and use these together with probability factors to find a realistic value, the expected commercial value of ECV.

The formula given by Cooper[8] is:

$$\mathrm{ECV} = [(\mathrm{PV} \times P_{CS} - C) \times P_{TS} - D]$$

where PV is the present value,

P_{CS} is the probability of commercial success assuming technical success,
C is the cost of commercialisation (product launch),
P_{TS} is the probability of technical success, and
D is the development cost remaining in the project.

The establishment of probability factors requires objective study of the past performance of the company, the competitive environment, and the level of innovation, which indicates technical risk. Techniques such as Failure Modes and Effects Analysis can also be used to assess technical risk.

Notional royalty method

This method of valuation uses the principles of licensing. Briefly, a technology is valued by imagining that it is being licensed in an arm's length transaction to another company. You then apply licence valuation principles to estimate the cost of the licence and calculate the present value of the licence fees over time. This gives a value for the technology. The valuation of licences is dealt with in Chapters 12 and 13.

The Toyota Prius, capturing the intellectual assets

The Toyota Prius is a hybrid vehicle with two engines, a 1.5-litre petrol engine and a 50 kW electric motor. It also carries large capacity nickel-metal hydride (Ni-MH) sealed batteries. In normal driving the petrol engine provides the power and charges the batteries. In urban driving the petrol engine cuts out and the batteries supply power to the electric motor thus cutting down on emissions and fuel consumption because the petrol engine is switched off in traffic.

The project was started by Toyota in 1993 and the first Prius was launched in Japan in 1997. The car has achieved numerous awards for 'Car of the Year' and 'Most Environmentally Friendly Car' in many countries. Over 500 000 have been sold worldwide up to end of 2005. The Prius produces only 104 grams of carbon dioxide per kilometre travelled compared with 170 grams for the average non-hybrid cars.[9]

A search of the patent database reveals 802 patents for hybrid-related inventions granted to Toyota up to April 2006. Toyota now effectively controls the hybrid vehicle intellectual assets and has licensed, Ford, Nissan, General Motors and Daimler-Chrysler.

Notes

1. Edvinsson, L. and Malone, M.S. (1997) *Intellectual Capital: The Proven Way to Establish your Company's Real Value by Measuring its Brainpower*. London: Piatkus.
2. Ibid.
3. See *http://www.ipambestpractices.com/index.html*
4. Petrash, G. (2002). *Greater Value through Intellectual Asset Management (IAM)*. Delphion.
5. Ibid. p. 9.
6. Dietz, M.C. and Elton, J.J. (2004) Getting more from intellectual property. *McKinsey Quarterly* (4).

7. Germeraad, P. (1999) Intellectual property in a time of change. *Research Technology Management* 42 (6): 34–39.
8. Cooper, R.G. (2001) *Winning at New Products*. Perseus, pp. 226–229.
9. Source: *Observer* Special report 12/03/06.

Contractual issues

The technology manager is faced with many issues that have a legal dimension. There are many relationships to be established between various parties involved in the introduction of new technology. The basis for these relationships, the obligations and performance of the parties involved, must be laid down in advance and where necessary, in the form of legal agreements. There can be many misunderstandings between people or organisations. The potential for misunderstanding must be identified in advance and, where possible, agreements entered into that specify what will happen in different scenarios so as to provide a remedy if things go wrong, or to make sure that rewards go in the right direction if things go right. Failure to face issues or draw up suitable contracts can result in conflict and financial loss.

The treatment of legal issues in this and subsequent chapters is written for the technology manager who is not expected to be able to function as a lawyer but nevertheless should be able to understand the legal implications of actions planned or taken and know when to call for legal advice from a professional.

Types of contractual issues

The types of contractual issues that a technology manager may have to deal with are linked to the different technology acquisition and commercialisation strategies that a company may adopt. In developing its own technology, a company must deal with inventors and issues of ownership of the technology. It must make sure, for example, that patent rights are correctly filed and that the necessary documents are signed by inventors. Frequently, development work is contracted out to universities or consultants, and the issues of ownership of their work, payment for services and liability if things go wrong must be addressed. Other issues that frequently arise concern the disclosure of confidential information or joint development projects.

A company that is seeking to acquire technology externally will need evaluation agreements to allow it to evaluate new technology. It will probably enter into licence contracts or may be involved in the purchase of companies. In commercialising its technology it may license it out to other organisations.

Principles of contract law

The most common form of legal contract is a contract for sale. A person agrees to sell something to another person for a certain price. The people who are dealing with each other are called parties. A company or an organisation can also be a party. The price that the parties agree is called the consideration. Consideration means something that is given or done in return for something else and does not have to be in the form of money. In legal Latin terminology consideration is called a 'quid pro quo'. Consideration can be an act, but it can also be a forbearance. In a patent licence agreement, for example, the licensor or patent holder agrees not to sue the licensee for infringement of the patent, provided the licensee pays the royalties or other licence fees specified.

Any form of sale involves an offer and acceptance. A product or service is offered for sale at a certain price. This is called an 'invitation to treat' or an invitation to discuss the price or specification if it is negotiable. Once the price and description of the goods or service are agreed the offer has been accepted. The contract is not completed until the consideration is paid and the product is delivered or the service provided.

A contract can be verbal or in writing. A verbal contract is no less valid than a written one but there can be problems of proving what was agreed if no written record exists. The courts will always seek to find out what were the intentions of the parties and it is only if this cannot be determined that a contract would be considered void due to uncertainty.

The agreement between parties and the completion of the contract may not necessarily be simultaneous. This gives rise to extra clauses in an agreement to define the timing of the execution of the deal and schedule of payments. For example in buying a house, the purchaser pays a deposit at a certain stage, then makes a further payment when the roof goes on. There is a payment on closing the sale and handover of the keys but there may also be a retention of monies until the purchaser is satisfied that the house is of the quality specified.

Although there may be timings of payments, the law is very strict on the subject of consideration. 'Past consideration is no consideration', is a well-known phrase. In other words you cannot enter into a contract now on the basis of something that was done or paid previously.

Frequently, in entering a contract for large amounts of money, there is an evaluation or investigation phase while the purchaser checks out various details. The purchaser may ask for an option or for first refusal. An option is useful if the price can be agreed subject to other factors such as the availability of finance or planning permission. With an option you agree to negotiate now, and have the option to complete or not complete the deal within a fixed period, say 6 months. The purchaser pays a price for the option to compensate the seller for the fact that property or goods cannot be offered for sale to any other party during this time.

In the purchasing of technology, options are not useful when the purchaser is evaluating the quality of the technology and cannot agree a price until the evaluation is complete. A request for first refusal is more useful in such a case. This means that the seller must give the purchaser the chance to match any other offer the seller may get during the evaluation period.

Contracts also deal with issues such as non-performance by either party and may specify remedies to be taken, or financial penalties. If this fails, either party may go to court for an 'order for specific performance' and the court may instruct the party to fulfil its obligations. Contracts also deal with more general issues and contain a lot of standard clauses. Lawyers call these 'boilerplate' and they deal with all the 'what happens if' scenarios. Examples of these scenarios are that either party might become insolvent or be taken over or there might be a war.

A contract for sale ends on completion of the sale, but other contracts can last for long periods and so the duration of the contract must be specified. If a company agreed to provide a service at a fixed price over a number of years, they might regret that they did not check the duration of the contract if their costs had increased. Provisions for termination may also be included so that either side can get out of the contract if it finds that the arrangement is no longer in its best interests.

As the law of each country is different, the contract should specify which country's law is applicable. This is very important in a contract between parties in different countries. For example, signing a contract under US law with a US company without understanding the implications of US law could be an expensive mistake.

R&D contracts

In R&D contracts, a service is offered and payments are specified. Issues that need to be addressed in detail are the description of the work programme, payments and intellectual property (IP) matters. The work programme must specify exactly what work is to be undertaken and the results expected. Specific technical performance targets may be set and these are called the milestones and deliverables. Milestones are important stage completions at which the progress of the project can be assessed. Deliverables are physical items that could include prototypes, software, reports, tooling or equipment.

There may be stage payments as each milestone is reached and the contractor may retain title to IP or deliverables if payment is not made in the specified time after completion. The R&D contractor is usually asked to agree to assign any IP rights arising out of the work to the party commissioning the contract. The issue of ownership of IP will be dealt with in more detail below.

Funding contracts

Many R&D programmes are funded by government agencies or by the European Commission. The funding agency sets down conditions under which its funding is made available. In general, these contracts specify the following:

- The description of the work.
- The timing and milestones.
- Requirements for commercialisation or dissemination of results.
- Amounts of funding from the agency.

- Arrangements for drawing down funds.

Funding contracts with the European Commission usually specify a number of points in addition to the above. Because EU programmes are multipartner and transnational, the partners are specified and the co-ordinator who acts on behalf of the others in dealing with the Commission is named. The duties of the co-ordinator in communicating with all the parties and in disbursing the monies are specified. There are also provisions on ownership of IP and dissemination of results or making the results available to the parties and to others on commercial terms.

Licence contracts

A licence means a permission and so a licence contract gives a party permission to do something it was not permitted or able to do before. The permission is granted by the other party based on their IP, which may be in the form of statutory rights such as patents or copyrights, or non-statutory rights such as know-how.

The main provisions of a licence contract describe the rights or know-how to be licensed, the consideration, a downpayment on signing or a royalty on sales or both. The licence also specifies the territory covered and whether the licence is exclusive or non-exclusive. As well as giving permission to use restricted information, the contract may specify how technology is to be transferred from the licensor to the licensee. Other clauses cover a number of 'what happens if' scenarios and the steps to be taken if they occur. These would include release of confidential information, infringement of IP, technology transfer and technical assistance. These points will be dealt with in detail in Chapters 12 and 13.

Software contracts

A lot of software contracts are special forms of licences. In purchasing mass-market software you have to accept licence terms by clicking your mouse or by opening the package. This type of licence is called a 'shrinkwrap' licence. The name arises from the fact that you are supposed to read the licence terms through the transparent packaging. The purpose of this type of licence is to prevent the purchaser from making more than one copy of the software or giving copies to others. When the software is to be used on a number of machines or on a network, the type of licence is called a site licence and specifies the number of machines and type of payments.

Other contracts cover such matters as permission to act as a distributor, to amend or repair software, or to convert it to a different language or operating environment. There are contracts for maintenance services specifying the type of assistance, on-line, telephone, or on site, and the commercial terms for each.

Purchasers of specialist software need to be protected if the developer of the software ceases to trade or refuses to provide a maintenance or upgrade service. This protection is provided by an escrow agreement. A copy of the source code is put into safe keeping with a third party called an escrow agent. In the event of the software developer failing to meet its obligations the purchaser may apply to the escrow agent for the release of the source code so that the software may be repaired or upgraded. It is important to make sure that the developer fulfils any obligations to keep the source code up to date.

Ownership of intellectual property

If a piece of IP is created by a person, they are automatically the owner of it except where that person is an employee. Work done by an employee in the normal course of his or her duties is considered to be the property of the employer. This arises from the old 'master and servant' relationship in common law where the master is considered to own the work done by the servant.

Patents

The legal position in relation to the ownership of patents was decided by an important court case in the UK. A storeman who was an employee of Electrolux invented an improvement for a vacuum cleaner and applied for a patent in his own name. Electrolux challenged the application stating that he should have given the patent to his employer. The case went all the way to the House of Lords who decided in favour of the inventor because he was not expected to invent things in the normal course of his duties as a storekeeper.

However, anybody in a technical or more senior capacity, or one who, in the words of the judgement, is in a position where he or she is expected 'to further the interests of his employer', would be considered to be inventing something on behalf of the employer who would automatically be the owner.

A grey area arises where the invention is made in the employee's spare time. Mostly, such inventions are related to the employer's business or use knowledge or facilities provided by the employer. In such cases, the employer would most likely be the owner. Where the invention was not related to the employer's business or was in a completely different area of technology, the employee would own it.

When a patent application is filed by a company, the employee inventor is required to sign a deed of assignment transferring ownership formally to the employer.

Joint ownership of patents

Where patents are jointly owned by more than one person or organisation, each owner has full rights to exploit the patents independently of the other(s). It is essential that the joint owners enter into an exploitation agreement that specifies that they will act together on all matters concerning the exploitation of the patents. Failure to do this can result in disputes leading to loss of exclusivity in licensing or marketing or can even threaten the commercial success of a project.

Joint projects between a number of partners such as those carried out under the EU Framework Programmes usually result in IP being owned by the partner who develops it. However, where two or more partners are working together on a particular problem, joint inventing and joint ownership can result. Usually the partners are required to enter into agreements to exploit the IP by agreement rather than by independent actions.

Joint projects bring up complex issues concerning access to pre-existing know-how that may be required by the partners to carry out the work. The project may result in the creation of new IP that the partners may wish to exploit and may need access to the know-how or newly created IP of other partners to do this.

University inventions

Problems can arise in projects with universities where postgraduate or other students are working on an R&D project on behalf of a company. As these people are not employees of the university, their inventions belong to them and not to the college. It is essential to ensure that all individuals working on such a project sign contracts agreeing to assign any patents resulting from their work on the project to the university or company commissioning the work as may be required by the R&D contract.

If the work is paid for by the company with no contribution by the college then any ensuing patent rights belong to the company. If, however, the college decides to become involved by providing facilities or researchers then the college becomes part owner of any patents. Before the start of the project the payment in kind by the college must be valued and a percentage sharing of the future patent rights agreed. This should take account of which party is to pay for the filing and obtaining of foreign patents. In general, joint ownership of patents can lead to difficulties in decision-making and to complex administrative problems.

Copyright

The position for copyright is similar in practice to that for patents. The author is considered to be the first owner of copyright but where the author is an employee and the work is written for the employer, the employer is the owner. This is important for authors of software, graphic designs or other copyright works used in business. However, universities are usually quite happy to allow authors who are employees to publish papers and books and to own the copyright.

Where a copyright work is commissioned, the work itself belongs to the commissioner, but the ownership of the copyright in the work is the subject of negotiation between the parties. If agreed to belong to the commissioner, it must be assigned by a deed of assignment before the commissioner becomes the owner and has full freedom to use it. Many problems have arisen in the area of graphic design where designers have produced works for commercial organisations for advertisements. In most cases copyright is not assigned, yet advertisers have continued to use the copyright works or to modify them or use them in later campaigns without adequate compensation to the designers.

Problems can also arise in the licensing of software from the developers to the end-users. The licensing of software will be dealt with in Chapter 12.

Dealing with inventors

In dealing with inventors the first question to determine is whether the inventor is an employee. If the inventor is an employee of your organisation you will have to make sure that the invention is correctly communicated to your patent agent and that a deed of assignment is signed. You then have to ensure that the inventor is available at a later stage to assist in dealings with patent offices concerning the prosecution of the patent applications through the patent examination process.

Many organisations have different kinds of reward schemes for inventors, which may involve arrangements to share in royalty from inventions. These can involve complex issues concerning tax or the calculation of the percentage of royalty or other compensation to the inventors.

If the inventor is an employee of another organisation you must make sure that the inventor is acting on behalf of that organisation and not on his or her own behalf. If the inventor claims that the invention does not belong to his employer it would be normal and prudent to get a letter from his employer waiving any rights to the invention before you could enter discussions to take it over or license it from the inventor.

Inventors often approach companies offering their inventions for license or requesting some form of joint development cooperation. Very few of these approaches result in any form of cooperative agreement or in commercial success. Possible reasons why this is a problem area are:

- Many companies are not interested in outside submissions and don't take them seriously.
- Many inventions by private inventors have been shown to be of poor quality.
- Even professional inventors can be inflexible about how their invention is to be developed.

Inventors often have unrealistic expectations, failing to understand the size of the investment in development and commercialisation needed before any money is earned.

It is essential to deal courteously and speedily with any outside invention submissions received. A company should have a procedure for processing internal inventions and this should be used to carry out an evaluation of any external submissions.

Disclosure and evaluation agreements

It happens very often that companies and individuals need to disclose confidential information to each other but must be sure that before this information is disclosed that it will not be disclosed to others or used to the detriment of the discloser. These agreements are called non-disclosure agreements (NDAs), secrecy agreements or confidentiality agreements. Possible reasons for using such agreements are:

- Discussing information on inventions before patent applications have been filed.
- Evaluating a project or invention submitted by an individual or company to another company.
- Approaching an investor to invest in a project.
- Evaluating the possible purchase of a company.

Non-disclosure agreements

An NDA is drawn up between two parties. These are called the discloser and the recipient. The principal features of a NDA concern the identification of the material to be disclosed, the agreement by the recipient to keep the information confidential and the conditions under which the recipient is released from any obligations entered into.

The recipient may be an individual but may be representing a company. If the information is being disclosed to the individual who is then going to discuss it with colleagues in the company, it is important that the agreement binds the company and that the individual is a person who has authority to sign and enter undertakings that bind the company.

Lawyers often insist on very wide definitions of what is being disclosed. This can have the effect of making the recipient feel that he or she may be prevented from working on something that they already had knowledge of, so it is important to state that the recipient is not bound in relation to information they already possess. It is also normal to state that the recipient is not bound to confidentiality in relation to information disclosed to them that subsequently enters the public domain either by disclosure of the discloser or from any other source.

NDAs should be as simple as possible so that executives can sign them without feeling that they need to refer them first to their legal advisers. There are other much more complex forms available that include non-competition clauses and threaten serious consequences for breach of the agreement. This type of agreement can produce serious delays in the disclosure and evaluation of information and is unlikely to be any more effective. In many cases the remedies for breach of such agreements can rarely compensate for the possible loss of patent rights or other damage to the company's business interests and therefore such agreements should be regarded as a discipline and a warning and one would hope never to have to go to court.

The duration of such agreements should not bind the parties to keep the information confidential indefinitely as this may be an unreasonable request. NDAs can also be drawn up in a mutual form where both parties disclose information to each other and agree to keep the other's information secret.

Evaluation agreements

Evaluation agreement are usually combined with NDAs or sometimes with agreements covering the delivery of specific material such as biological samples, software or other material or prototype products for test. These agreements usually specify the type of evaluation to be carried out, the duration of the evaluation and whether the outcome of the evaluation is to be reported to the discloser in detail giving reasons for any decisions taken.

An example of such an agreement would be a case where a university department might provide samples of a genetically modified vaccine to a pharmaceutical company for evaluation. The university would agree the duration and nature of the tests. The university would require details of the performance of its samples and might ask for permission to publish such information or be free to release it in confidence to others should the pharmaceutical company not be interested in the project. They may also require the return of the samples and any other products derived from them.

12

Licensing and technology transfer

Introduction

Licensing, and the more general term 'technology transfer' (TT), are technology acquisition strategies that are concerned with the sourcing of technology externally. Chapter 3 covered the subject of identifying technological opportunities. This chapter deals with the mechanisms for obtaining the technology and covers issues such as the forms of TT', getting permission to use someone else's technology, licensing principles and methods of payment for technology.

Definitions

There are many terms used in the area of TT. Some of them overlap and some are often misunderstood, so it is worthwhile to give some definitions.

Technology transfer

A general term covering the movement of applied science from one organisation or country to another. It can take many forms, including licensing, imitation, purchase of equipment, recruitment of specialists, and purchase of businesses.

Licensing

Licensing is the granting of rights to carry out a restricted act. Industrial licensing usually means granting permission to manufacture and sell a product or to use a process. The subject matter of the licence may be covered by intellectual property (IP) rights such as patents or may be in the form of trade secrets. Royalties are usually paid to the licensor by the licensee. After the granting of a licence, a TT phase usually takes place where the licensee is instructed in the use of the licensed technology. Licensing-in is the activity of the licensee or recipient and licensing-out is what is done by the licensor or owner of the technology.

Joint venture

A joint venture involves a number of partners setting up a new jointly owned company to exploit a business opportunity in a certain country. It often includes a licence from one partner to the joint venture company. The other partners would usually contribute finance or market access.

Technology scouting

Technology scouting is the name applied to carrying out a search for new products, processes or technologies. The term includes product and technology searching. Technology scouting can be done to meet the needs of a specific company or on a speculative basis to identify strategically useful technologies.

Forms of licensing and technology transfer

The general term TT means simply the transferring of technology from one organisation to another. It does not specify the mechanism involved. Different mechanisms involve different issues, some legal, some commercial. Assuming the source of technology has been identified, we can then discuss how the technology is transferred.

Purchase of equipment

One of the most widespread methods of TT is the purchase of equipment. An example of this would be a company that manufactured timber products using manually controlled saws and routers. This company then purchases computer numerically controlled equipment. Included with the equipment would be a training programme for operatives to enable them to program the machines. The company has now upgraded its production technology and is able to move towards computer-aided manufacturing (CAM) where computer-aided designs can be transferred directly to production.

This type of TT does not require any transfer of rights. It is a simple technology purchase with a single payment, although there would probably be a site licence for the software. The company acquires technology on a non-exclusive basis. The seller of the machinery is free to sell to their competitors.

Licensing

The word 'licence' means permission, and it is that element that distinguishes this type of TT from the previous one. The permission may be to use a patent or it may be to have access to secret know-how. In general here we are talking about industrial technology. The subject of software licensing is somewhat different, and will be considered separately.

The licence may be to make a product. If this is the case, it means that the licensor, the developer of the technology, will be allowing someone else to make their products. They have to decide whether the licensee, i.e. the firm that obtains the licence, will be allowed to compete with them in their own market or not. This is a completely different situation to TT by purchasing technology, because here the licensor is giving away some of their market or potential market.

If the licence is for a process, it may also allow the licensee to compete with the licensor's products, or if the process is applied in a different industry there may be no competition at all. An example of this is the development of ink-jet printing, which was originally developed for printing date information on fast-moving goods on a filling line and was then adapted for printing on paper. This is sometimes called spin-off technology.

It can be seen that in licensing its technology, a company is giving away some or all of its access to a potential market, and so it seeks to get a payment to compensate it for not being able to operate in that market.

Recruitment of specialists

Firms often acquire technology by recruiting technical specialists who have the knowledge that the firm requires. This assumes that the specialists are free to use that knowledge and there are many cases where scientists and engineers are poached from competitors and bring significant amounts of know-how with them. Employment contracts seek to prevent employees from releasing information to competitors when changing jobs. Employees are bound by confidentiality and non-competition clauses but these can be hard to enforce. TT by employees changing jobs is often called 'TT on the hoof'.

Imitation

Technology is also transferred by using information that is obtained legitimately. Companies buy their competitors' products and take them apart and find out how they work and then develop their own similar products. This is called reverse engineering. Information is also obtained from technical and patent literature and provided there is no infringement of IP, this activity can be carried on and is engaged in quite often. It is usually an opportunity to improve on the competitors' products, which is a better strategy than copying and less likely to lead to threats of legal action.

Purchase of companies

Larger companies often acquire technology by taking over smaller ones. This happens when larger companies are threatened by new technology coming from smaller and more innovative companies that are weaker financially. Sometimes the smaller companies have run out of cash having spent a lot of money on R&D and are ripe for takeover.

Licensing principles

Licensing is an activity that requires the cooperation of two parties, the licensor and the licensee, whereas some of the other forms of TT are more one-sided. The principles behind licensing are mainly concerned with the relationship between the licensor and licensee. As licensing requires the licensor to forgo some of its operating freedom, it is worthwhile to list the different types of licences and how they restrict the licensor's freedom to manoeuvre.

Exclusive licence

In granting an exclusive licence, the licensor gives permission to the licensee to operate the technology and undertakes not to compete against the licensee and not to grant a licence to any other party.

Sole licence

A sole licence differs in that the licensor reserves the right to compete in the market with the licensee but cannot grant any further licences.

Non-exclusive licence

The licensor is free to compete with the licensee and grant any number of licences.

Field of use

Here the licensor gives away rights in some markets or fields of use while retaining others or licensing others to different licensees. A company that has developed a concrete moulding technology might license one company to use it to make houses but license a separate company for the industrial applications.

Territory

The territory is the area, country or countries covered by the licence agreement. The territory is often dictated by the IP coverage. Exclusive licences are often granted for a certain territory.

Profit-sharing model

The financial principles behind licensing are based on a profit-sharing model. Imagine that the developer of the technology decided to exploit all the possible markets itself by manufacturing and selling products using its patented technology. It then decides that some markets are very distant and that it could take a long time to build up a presence there. It now decides to seek a licensee in that market and works out the size of the market and gets an indication of price levels.

When it negotiates with a licensee, it considers the profit it would have made by exploiting the market itself, and the risk factors involved. Generally, once the technology has been developed there are few risks in licensing so it is considered that the licensee will be taking a bigger risk than the licensor. The licensee must set up to make the products and then sell them at a profit. For taking the bigger risk the licensee expects to make a profit on manufacturing, which is bigger than that made on licensing by the licensor.

Royalties

One of the fundamental principles of licensing is that payment for the licence is made in the form of royalties, a type of rental or usage payment. The most common royalties are based on sales, i.e. a percentage of sales revenue is paid to the licensor by the licensee. Royalties can also be charged per unit sold, or on a usage or throughput basis in process technology.

The licensor is not concerned directly with the profitability of the licensee. It is up to the licensee to make a profit and the licensor receives the same royalty on sales regardless of whether the licensee makes a profit or not. The licensee can price the products as required and obviously a higher price means more royalty income to the licensor if the royalty is based on sales revenue. Conversely, a higher price may result in a lower volume of sales and so the total royalty may fall. Once the licensor has signed the licence contract these factors have been ceded to the licensee and the licensor is in the position of sitting on the sidelines and collecting royalty at intervals but not interfering except when asked for assistance.

Rights granted

IP rights cover the making, using or selling of a product or service and licence rights may be separated out on that basis. A licensee may be given rights to manufacture in one location or the use a product in a certain territory. A sales territory may also be specified. Territories are usually linked to IP coverage and it is normal for a licensee to pay a higher royalty to sell into a territory where it has the benefit of patent protection.

Know-how licences

The use of confidential information or secret know-how does not involve legal rights like the right to use a patent, so know-how licences differ from patent licences. In a know-how licence the licensor agrees to disclose secret information to the licensee and the licensee agrees to preserve the secrecy of the information obtained as if it was their own. Royalties are usually paid as in a patent licence but the rate is lower because no patent protection is passed on to the licensee.

In general, know-how licences run for shorter periods than other licences. This is because secret information tends to leak out into the public domain and a licensee does not want to pay for something that becomes available freely to others.

Franchises

Franchises are usually based on a combined trade mark and know-how licence. The most famous franchise is McDonalds, which by 2002 had grown so large that there were four new outlets opening every day worldwide. This is sometimes called a business format franchise. It means that what is being licensed is the format or winning formula of a certain type of business.

Franchising is suitable for businesses that can easily be replicated if you have the information. There is usually an operating manual that specifies exactly how to run the business down to the last detail. This, together with advice on start-up, equipment and training, is the know-how. This is usually supported by a strong brand identity backed up by trade marks and national or even international promotion.

Franchisors usually recruit candidates from entrepreneurs or redundant managers and provide access to special start-up financial packages from the banks. They usually look for people who know how to run a business but are not expert in the business of the franchisor. They believe that for a franchise to work, the franchisee must need them and be prepared to do things exactly as in the operating manual. Turning a business into a franchise can be a very successful business expansion strategy, but it is only suitable for businesses that serve local markets and are easily documented and packaged.

Technology brokers and consultants

Brokers are commission agents who arrange deals and obtain their payment in the form of a commission or percentage of the sale amount. They are not a party to the sale contract and are not responsible for the performance of the seller. They make representations on behalf of the seller as his agent but the purchaser knows that he or she is dealing with the principals. In TT, brokers always work for the seller of the technology.

Consultants usually work on a fee basis. They can represent either the buyer or the seller of the technology and provide services such as product and technology searching, project or licence evaluation, and negotiation and contracts.

There are only a small number of successful consultants in the TT area in Europe. Technology searching on a no-foal no-fee basis has not yet become a viable business and most consultants work on a fee basis or a mixture of fees and commission or royalty. Usually, they are commissioned by a potential licensee to search for technology and no fee is paid by the licensor.

There are more successful firms in the USA due to the higher level of TT activity there both inside the USA and internationally. US consultants can provide excellent on-the-ground knowledge to identify suitable technology. Technology brokerage is also being carried on in the Far East.

Payments for licences

Licence agreements usually specify a downpayment or up-front fee on signing and a royalty on sales. The downpayment reflects the size of the opportunity and territory and may include the costs of transferring the technology. Royalties reflect the margins or profitability of using the technology. The calculation of these payments, and how much a licensor would expect or a licensee should pay will be discussed in Chapter 13, which deals with TT negotiations.

Some licences are described as 'fully paid up'. This means that there is an up-front fee and no royalties. This type of licence is used for a limited TT know-how licence or for software licences.

Exclusive licences have minimum royalty clauses. In return for being the only licensee in a territory the licensee is expected to perform and to achieve a minimum level of sales. The penalty for not reaching the agreed sales level is to pay the royalty appropriate for that sales level. The agreement may also provide for converting to a non-exclusive licence, so that the licensor is not stuck with a poorly performing licensee.

Royalties can also be structured on sliding scales. In some industries, because of long and expensive start-ups, the licensee may request a lower royalty rate in the earlier years, gradually rising as they become more profitable. In other industries the royalty may fall over time if the licensee has successfully argued that the licensor's take in later years would be excessive.

If a licence does not work out, sales will be disappointing and the licensee can approach the licensor and ask to terminate the contract. Licensing has the advantage that you only pay for actual usage of the technology and if it does not work out, it is relatively easy to disengage.

Software licensing

Software licensing was discussed in general terms in Chapter 11. Most software licences are actually software sales but they are arranged as licences so as to protect the IP rights of the developer especially where there is no contract between the developer and the end-user.

Software licences approach closer to the normal model of licensing where the following conditions, or some of them, apply:

- There is a contractual relationship between the developer and the end-user.

- There is an ongoing relationship which continues after the initial supply of product.
- The software controls an industrial or business process.
- Ownership of IP remains with the developer.

In software development contracts, the developer may retain ownership of the copyright or other IP in the software. This is different to the position that was described for R&D contracts in Chapter 11. The ownership of copyright will usually depend on the relative negotiating strengths of the parties.

If the commissioner of the development contract is strong and the software is being developed from scratch to their exact requirements, it is usual for the commissioner to insist on owning the copyright. If, on the other hand, the commissioner is not so strong and the developer is well established or is using an existing platform from which to develop an application for the commissioner, the copyright will usually remain with the developer.

In the first case, the copyright will be assigned to the commissioner on delivery of the software and a licence will not arise. In the second case, as the IP remains with the developer, a licence agreement will be necessary. The licence agreement is usually a combination of a copyright licence and a secrecy agreement preventing the licensee from making unauthorised copies of the software. Other issues, as laid out in Chapter 11, cover access to source code, and provisions for maintenance and upgrades.

EU licensing regulations

Articles 85–90 of the Treaty of Rome lay down competition rules that have a bearing on the enforcement and licensing of IP rights and other issues such as joint R&D projects and distribution agreements.

The purpose of the articles in the Treaty is to ensure the free movement of goods within the single market. A patent holder could license one manufacturer for the UK and another for Ireland and could use the fact that patents cover separate countries to prevent competition between the manufacturers in Ireland and the UK. If the manufacturer in Ireland tried to sell into the UK, they could be prevented under patent law, but this could contravene Community law.

This subject is complex and further reading is suggested if required.[1] An attempt will be made here to summarise the Community regulations as they refer to patent licences.

A general obligation exists to notify the Commission of certain types of agreements. However, an exemption is granted for agreements provided they do not contain blacklisted clauses. This is called the Block Exemption Regulation for patent licences. Briefly, licence agreements may not:

- Prevent the licensee challenging the validity of the patents.
- Extend the licence term beyond the life of the patents.
- Apply limits to customers, prices or quantities of products manufactured or sold.
- Prevent competition with other licensees more than five years after the product was first marketed.
- Require the licensee to hand over rights to improvements to the licensor.

In general, the thrust of the regulations is to encourage the granting of licences for the whole of the EU. Where licences are granted only for certain Member States, thereby carving up the single market, licensees may be prevented from actively selling into another Member State for a period of 5 years, but passive selling is allowed.

The Yamaha DX-7: popular music from Stanford

The Yamaha DX-7 synthesiser has been described as the most popular musical instrument of all time.[2] What most people do not know is that it was based on technology developed by Stanford University.

In 1977 US Patent 4,018,121 was granted to Professor John Chowning of the Center for Computer Research in Music and Acoustics at Stanford University. Equivalent patents were granted in the UK, Germany, France, the Netherlands and Japan. While the patent was pending, the Stanford Office of Technology Licensing offered licences to a number of US corporations all of whom turned it down.

The College then made a cold call to the Los Angeles office of Yamaha and by a stroke of good luck Mr Kazukiyo Ishimura was visiting from Japan and was introduced to Professor Chowning. Thus began a long association that has outlived the patent and continued on to include later technology developed in Stanford.

US patent 4,018,121 claims a method of synthesis of sounds using frequency modulation at audio frequencies. The original application was for an electronic organ but this was later widened to a musical instrument. Yamaha put considerable development effort into the project and it took them 10 years before they had produced the DX-7. Sales increased rapidly worldwide and the Yamaha licence became the second largest royalty contributor to Stanford's licence income, which by 1995 had reached $38.2 million.

Ironically, once the product became well known in the industry, Stanford began to receive complaints from US corporations that the technology had been licensed exclusively to a Japanese company.

Notes

1. Byrne, N. (1994) *Licensing Technology: Drafting and Negotiating Agreements*. Macmillan.
2. Koepnick, J. (1995) Case study: Stanford's FM License. *Les Nouvelles, Journal of the Licensing Executives Society* 30(2): 87–90.

13

Technology transfer negotiations

Introduction

Technology transfer negotiations take place in the case of a licence or sometimes as once-off technology purchase. This chapter considers the commercial and contractual issues as well as the importance of seeing licences as strategic alliances where relationships are built up between firms, which are of mutual benefit to both sides.

Technology offers

Developers and owners of technology engage in licensing either on an active or passive basis. In many cases it is passive. The licensor is approached by a potential licensee as a result of a technology search and the licensor may not have an active licensing programme in place. In this case the licensor must decide on the basis of the approach whether they are prepared to grant a licence or not. Factors to be considered will be the presence or intended presence of the licensor in the market or territory requested, the perceived ability of the licensee to exploit the market in a reasonable timescale and the strategic implications of a partnership with the potential licensee.

Other licensors may have an active licensing programme. These may include bodies that do not manufacture, such as universities or research organisations, or companies that have decided to exploit only certain markets through manufacturing. Organisations that engage in licensing on an active basis usually make their technology available through licence offers, which are publicised through various marketing channels. The sources of information on such licence offers have been described in Chapter 3.

An initial technology offer document is usually quite short and has the purpose of informing the potential licensee of the basic information about the opportunity so that a decision can be taken whether to request further information or not. Thus the initial document is strong on promotion, on the benefits of the technology and the intellectual property (IP), but reveals nothing about prices or royalties.

Responding to technology offers

When a company has received a technology offer, either in response to a technology search or an active approach, it has to carry out an initial screening of the offer. The purpose of the initial screening as described in Chapter 3 is to decide if more information is needed or if the offer is not of interest on the basis of the information put forward.

Screening is done by setting 'must meet' criteria for the licence offer. These are similar to those used in project evaluation for R&D projects as described in Chapter 5. Screening should be done by a multidisciplinary team and involves evaluation without investigation. In other words, a screening committee considers the offers on the basis of the information in front of them and decides to reject those that don't meet the criteria and to seek further information on those that do.

In seeking further information it may be necessary to enter into non-disclosure agreements and to travel to the office or factory of the licensor for an initial meeting. At the same time, information should be sought from published sources to give a degree of objectivity or credibility to the information supplied by the licensor, or to verify certain points. This could even include purchasing the licensor's products and testing them.

Other pre-contract enquiries could include the normal type of 'due diligence' information of a legal and accounting nature, which is sought in entering into an agreement. This could include verifying that the licensor was the owner of the IP and had the freedom to grant the licence sought, checking on the claims of market coverage and looking at published accounts.

Licence evaluation

The evaluation of licence opportunities is similar to normal project evaluation, but with some extra dimensions. In evaluating a project for introducing a new product, one would normally expect to have a sales forecast based on market research that would project for 3 or 4 years ahead the future sales volume and value. From the information on manufacturing the product it would be possible to project the costs, and when the costs are subtracted from the revenues you would get a projected gross margin. From this gross margin you would deduct the marketing and administrative overheads and any start-up costs to arrive at a net margin.

In order to estimate the capital requirements of a project it would be necessary to do a cash flow adding in the additional cash requirements for research and development, and plant and equipment. This calculation would generally show a negative cash flow in the first few years, becoming positive and showing increasing profits as time goes on. The maximum negative cash flow figure is the amount of money that must be invested in the project.

To arrive at the above financial model of a project, it is necessary to carry out a market evaluation in order to develop, with some confidence, a sales forecast based on a realistic selling price. It is also necessary to carry out a technical evaluation so as to be able to arrive at a production forecast and to give realistic estimates of production costs and investment requirements. In the case of a licensing project it is also necessary to investigate the strength of IP protection which is being offered to the licensee. Thus one could summarise a licence evaluation by saying that three aspects have to be looked at:

- Intellectual property.
- Technical evaluation.
- Market potential.

These three items must be considered together in order to decide what the licence is worth. This assessment is compared with the asking price in terms of downpayment or royalties, and the potential licensee is then in a position to decide whether to enter into negotiations or whether to consider alternative ways of getting into the same business.

Evaluation of patents and other intellectual property

A considerable amount of patent investigation can be done by desk research as described in Chapter 9. When the investigation begins to indicate a serious interest, it is wise to seek the aid of a patent agent. The evaluation will involve consideration of a number of aspects. It is possible to obtain definite information on these aspects fairly quickly and for this reason it is wise to evaluate IP first before spending a lot of time and money on some of the greyer areas of evaluation such as market research. Aspects of patents that should be investigated are:

- Scope of patent claims.
- Territory and duration of patent coverage.
- Validity of patents.
- Competing patents.

By carrying out patent searches, it is possible in conjunction with the patent agent to assess the scope or strength of patent coverage. This means that you look at the patent claims, i.e. the legal statements of what the patent covers, and assess how broad the patent coverage is. At the same time it is possible to assess the geographical coverage and determine the legal status of the patents in terms of which countries they are granted in. By carrying out patent searches to assess the novelty of the patents in relation to other competing inventions, it is possible to express views on the validity of the patents and to obtain information on other competing technologies, which may not have come to the attention of the potential licensee.

Evaluation of copyright, designs and trademarks is more limited. In these cases you are concerned more with questions of ownership or registration details. The actual content covered by these forms of IP would more properly be the subject of a technical or commercial investigation.

Many licences offer a significant amount of know-how and its value should not be underestimated. Know-how from manufacturers covers the detailed experience that the licensor company has obtained over the years in learning how to make its products efficiently and effectively, and contains information that may be copyright but would not be sufficiently inventive to be patentable. Some of this know-how may be disclosed to the potential licensee under a non-disclosure or secrecy agreement in order to enable an assessment to take place. Alternatively, it may be a question of assessing the value of the know-how on the basis of its end result without knowing how that result is achieved. The important thing in purchasing know-how is to make sure that you are not purchasing information that is available free of charge or at a small cost from published or other sources.

Copying

In the course of an investigation of a licence it may become clear that the technology on offer is not well protected and possibly that the know-how does not represent a significant advance on that which is already available in the licensee company. In this case, it is worthwhile considering whether it would be possible to design around the patent or other IP protection and to assess the costs and legal risks involved in this. Armed with this information, the company can then enter a licence negotiation in a stronger position, or alternatively, decide not to enter the negotiations at all and to develop its own product independently. There are considerable legal risks in going this route, and detailed advice from a patent agent including an assessment of the probability of being sued must be obtained. It should be borne in mind that the cost of litigation in this area is very high and would severely strain the financial resources of smaller companies.

Technical evaluation

There are a number of aspects to be considered by a company carrying out a technical evaluation of a licence opportunity. All have a bearing on the negotiating stance and may be crucial if the technology is new to the company. Consideration must be given to the following:

- State of development.
- Technical feasibility.
- Technical performance.
- Technical compatibility.
- Cost of manufacture.
- Cost of further development.

- Technical assistance.
- Future technology.

The state of development is the level of development sophistication of the product offered in relation to the 'state of the art', i.e. the general level in this area. The state of development also includes the question of how ready the product is for production. For example, if the product offered is only at the prototype stage, then a considerable investment must be made before the product can be put into production.

Technical feasibility refers mainly to inventions or products that have not yet gone into production. It is important to assess the technical performance of any product offered, and this should involve obtaining and testing samples. Where there are relevant standards, the products should be tested in relation to these.

The technology being licensed should also be compatible with the licensee. In other words, the licensee should have the capability of absorbing the technology without expecting any major unforeseen problems. If the technology involves a major change in technical sophistication, the potential licensee must recruit well qualified staff to absorb and embed the technology.

Detailed information should be supplied by the potential licensor on the costs of manufacture of the product including costs of machinery, materials, labour and overheads, etc. An assessment should also be carried out of the cost of further development including the cost of developing the product to comply with appropriate standards, for example, European electrical requirements.

Finally, the technical assistance capability of the licensor must be assessed. It may be necessary to call upon the licensor if problems are experienced. A licensor with a strong R&D capability will be able to provide this backup and will also be developing future products, which would be available for license, if the first project is successful. Licensing should be considered as a long-term relationship that could lead to a continuing transfer of technology, with the eventual hope that the licensee would develop its own level of technical sophistication and be able to grant back its own improvements to the licensor.

Commercial evaluation

The main aspect of the commercial evaluation of any product is market research. The company must estimate the sales that it is likely to achieve for the product and the sales estimates must be based on reliable information from the market-place. A reliable sales forecast giving estimates of price and volume for the product covering the potential market envisaged by the licensee is an essential requirement for the production of a business plan. If the sales forecast is inaccurate, all the downstream calculation of profitability or investment requirements can be very wide of the mark. Sensitivity analysis (see Chapter 10) should be carried out in order to assess the sensitivity of the project to different levels of sales and price.

The project evaluation should be drawn together by a detailed scoring process. The best way to do this is to have a score sheet where various aspects of the project can be rated on a score of 1 to 10 as described in Chapter 5. After these aspects have been evaluated, the project can be given an overall score and this provides a good degree of objectivity. It is frequently useful to provide weighting factors for various aspects of the evaluation in line with the company's opinion of the importance of these aspects in its decision-making process. Sound commercial judgement should be the final determining factor in making a decision to pursue a particular project. This would be heavily influenced by developing a relationship with the potential licensor as negotiations progress.

Downpayments and royalties

Payment for a licence is usually in the form of a downpayment payable on signing the licence contract and royalties payable at intervals of 3–6 months thereafter, based on sales. Contracts can sometimes involve other options. One option is to make the downpayment an advance of royalties. This means that once the downpayment has been paid no royalties are paid until that equivalent amount of royalties would have been achieved on the basis of sales. Licence contracts also contain minimum royalty clauses, which represent a commitment by the licensee to achieve a certain level of performance, usually in return for exclusivity.

Assessing downpayments

The downpayment reflects the value of the technology and can be considered as containing certain components that can be separately evaluated. One component of the downpayment will reflect the quality, strength and extent of the IP coverage, while another will reflect the quality of the technology or R&D investment and the cost of technology transfer. A third will reflect the extent of the commercial opportunity that this new product or process will offer to the licensee. The downpayment figure that is asked for in negotiations is usually determined subjectively, and there is no particular rule of thumb that is universally applied. The cost of the R&D is a factor that can be discussed, and the licensee can compare this with the money that they would have to pay if they were to take their own technology to the state of that offered by the licensor.

However, only a portion of the R&D costs are recovered in downpayments and the royalty element will contain an element of recovery of R&D costs for the licensor as well. In general, it is considered that the downpayment should not exceed 10 per cent of the R&D cost.

Assessing royalties

It is somewhat easier to work out what royalty to pay because royalties are based on margins. The licensor must be prepared to provide accurate production costs and information on the sales volume and price. From this the licensee can determine the likely profitability of the product and work out the net profit as a percentage of sales. The appropriate royalty to pay would normally be within the range of one-quarter to one-third of the net profit as a percentage of sales. Obviously the licensor will press for a figure of a third or more, and the licensee will tend to offer a quarter or less.

It must be remembered that royalties are calculated on the selling price of the licensee. If the licensee sells the product on to wholesalers, the royalty will be calculated on the ex-factory selling price. If, on the other hand, the licensee has its own integrated marketing structure right through to retailer level, the royalty will be calculated on the final selling price. This is because the royalty can only be based on audited sales figures that are extracted from the licensee's accounts. If they are based on any intermediate figures they may be distorted by transfer pricing between related companies. Clearly, a higher percentage royalty will be required on the ex-factory price than that which would be levied on the eventual sales price.

It is better to calculate royalties as a percentage of selling price than as a fixed amount per unit sold. In this way, the royalties are protected against inflation and one can avoid having to insert price variation clauses into the agreement. Royalties can also be traded off against downpayments in the negotiation. A company that has a strong financial position or has grants available to it may prefer to pay a higher downpayment in return for a lower royalty in the future. This will make it much more attractive to the licensor, as downpayments tend to be the major sticking point in negotiations. The reason for this is that a downpayment represents a commitment by the licensee before the project actually starts. Royalties, coming as they do from cashflow, are somewhat easier to pay.

There is no such thing as a typical royalty because all projects are different and therefore the margins reflect different circumstances. It has been found, though, that the margins for certain types of products tend to be similar, and from this one can derive some general guidelines that are shown in Table 13.1. High volume low margin products tend to have a fairly low royalty while low volume high margin products have a much higher royalty.

The 25 per cent rule

A useful starting-point in any setting or evaluation of royalty levels is the '25 per cent Rule'. This states that the licensee should pay a royalty rate equivalent to 25 per cent of the net profit on sales for a product. An example of this is given in Table 13.2.

It can be seen that the net profit on sales is 20 per cent and so 25 per cent of this is 5 per cent, which is the royalty suggested in this case. The 25 per cent rule is based on the profit-sharing model of licensing, which was mentioned in Chapter 12. The licensor is considered to share in the profits of the licensee but the licensee is considered to be taking the bigger risk as the licensor is not involved in the market-place and the royalty is a percentage on sales regardless of whether the licensee manages to make a profit.

Table 13.1 Percentage royalties based on ex-factory price

Type of product	Volume	Margin	Royalty range %
General consumer	High	Low	0.5–4.0
Specialist consumer/ general industrial	Medium	Medium	4.0–10.0
Specialist industrial	Low	High	10.0–15.0

Table 13.2 Typical product cost breakdown. The 25% rule suggests that the royalty should be 5% in this case

Sales revenue	100
Cost of sales	50
Gross margin	50
Overheads	30
Net margin	20
Royalty	5

The licensor is considered to have incurred the risk of developing the technology at the time when the decision was taken to engage in the development programme. Now that the technology is developed and ready for license it is considered that the licensee who has to invest in manufacturing and marketing has a larger risk.

Remember that although the 25 per cent rule establishes a royalty figure based on profit margins, the actual royalty paid is based on sales and it is up to the licensee to achieve the projected profit margins.

In general, in negotiations, the licensor seeks 30 per cent or more of the net profit as a royalty and the licensee offers 20 per cent or less. Negotiations determine the final figure based on the strength of the parties and the cases they can make.

Heads of Agreement

Licence negotiations should be conducted on the basis of a summary of the main commercial terms, which is known as a Heads of Agreement. This covers the main points in the licence contract as agreed between the principals of the two sides. It is best to avoid negotiating on the basis of a draft licence contract, as negotiations of this type quickly get bogged down in the small print. People become more concerned with the meaning of certain clauses than with the main commercial terms. It is important to remain focused on the main commercial issues such as downpayment and royalties. After these are agreed, the other less important issues and the drafting of a legal contract can proceed.

The licence contract generally defines the extent of the licence that is being offered and the payment conditions. It works on the principle that as long as royalties are paid the contract remains in force. If there is a failure to pay royalties the contract is automatically terminated. An example of the main points of a Heads of Agreement for a licence is given below.

1. Parties to the contract.
2. Technology/IP being licensed.
3. The territories covered.
4. Exclusive or non-exclusive.
5. Consideration: downpayments, royalties.
6. Minimum royalties (if exclusive).
7. Improvements.
8. Infringements.
9. Accounts.
10. Technology transfer.
11. Termination and duration.
12. Law of which country.

It is essential that the licensee should have professional advice before entering into a licence contract. The above points can then be considered in great detail and the licensee's attitude to them determined in advance. Professional advice should be obtained from firms that have carried out this type of work before. These may be experienced licensing consultants, or patent agents or solicitors who have a specialisation in IP law.

The Heads of Agreement sets out who the parties are. They must have power to enter into the undertakings in the agreement. The licensor must have ownership or licensing rights to the technology. The technology may be conveniently described in a patent specification or it may need to be described in separate detail if it covers know-how.

The document should set out the territory and whether the licence is exclusive or non-exclusive. Then the downpayment (if any) and royalties are specified. If the licence is exclusive, there will be a minimum royalty clause. This is usually based on an agreed sales forecast that is considered achievable. The minimum royalty figure may be set at a percentage of the sales forecast, which is regarded as unsatisfactory. If the sales do not reach this level, the licensee must pay that level of royalty or the agreement may become non-exclusive.

The other clauses are the 'what happens if' scenarios and deal with the actions to take should either side improve on the technology, how infringement of the IP are to be dealt with and the reporting of royalties.

Other issues cover the period and nature of technology transfer, the conditions under which the agreement may be terminated, the duration of the agreement and the legal jurisdiction that will apply.

Licence negotiations

Licence negotiations work best where both sides seek a 'win–win' outcome. If one side is too strong, the benefits to the other side will be disappointing and the project will not fulfil its potential. In general, the parties should get to know each other and build up a relationship of trust before sitting down to discuss commercial terms.

The following requirements are needed to carry out a successful negotiation:

- The licensee must have carried out an evaluation of the project and must be in possession of facts and figures in relation to the legal, technical and commercial aspects.
- Professional advisors should be on tap and should have been consulted in advance of the commencement of negotiations.
- Successful negotiation requires the building of a good relationship between licensor and licensee. If this is not yet developed, it would be better to delay the negotiations because they will be difficult to conduct in a climate where both sides are wary of each other.
- Tact and patience are also required as well as physical stamina.
- All contentious points must be dealt with at the negotiations, otherwise they will surface at a later stage where there are considerable financial implications, and will be very difficult to resolve.

The most contentious issue in negotiations is the downpayment. This is seen by the licensee as money taken out of the project that could otherwise have been used for the exploitation of the project. Perceptions of the value of the technology and its protection can vary considerably, and it is important to get the main contentious issues out of the way first.

Discussions on royalties can proceed smoothly if both sides are prepared to discuss the costings of manufacture or exploitation and agree that royalties will be related to projected margins. If neither side is prepared to discuss costings, royalty calculations are often based on numbers that cannot be justified by any objective criteria and this can lead to difficulties.

Licensing the Workmate

When Ron Hickman was eventually approached by Black & Decker (see Chapter 7 for the story so far), his product was already showing signs of commercial success and of achieving strong patent protection. He agreed a royalty of 3 per cent on sales, although he was advised to look for at least 5 per cent. However, Hickman's view was that the product would be successful and that it should be produced in large volumes. He did not want to saddle the product with a higher price that might prevent it getting into the high volume bracket. He also received substantial downpayments.

The UK and European royalties were split 1.5 per cent each for patents and 1.5 per cent for know-how and copyright. (The UK law at that time allowed for copyright in 3D from 2D drawings.) In order to get the patent royalty Hickman had to guarantee that strong patent claims would be granted. For the rest of the world the 3 per cent royalty was for patents only.

Because the licence was exclusive, minimum royalty provisions applied. Black & Decker also had an option to purchase the European patents and they exercised this in 1977. Hickman was also obliged to sue infringers, the cost of the actions being shared as well as the damages received. This was a most unusual term to agree as patent litigation can be very expensive and the outcome can sometimes revolve on a technicality.

As a result of the enormous success of the Workmate, Ron Hickman became a tax exile and made his domicile in Jersey where he set up business as a professional inventor. While he is listed as the inventor on 68 patents, he never had another success like the Workmate.

Note

1. Roos, M.J. (1983) A case history: 'Workmate'. *Les Nouvelles, Journal of the Licensing Executives Society* 18(2): 102–111.

14

Commercial exploitation issues

Introduction

This chapter deals with the final stages of external technology acquisition and covers the implementation of technology transfer. It also includes some additional issues in R&D management. These are the subjects of commercial exploitation strategies, including the licensing out of technology, enforcement of intellectual property (IP) rights and quasi-legal issues concerning product liability and standards, which are intimately linked with the development and exploitation.

Issues in technology transfer

Negotiating the deal in an external technology transfer is only the preparatory stage to making and selling a new product or introducing a new process. In this case, the conventional R&D phase has been replaced by a technology transfer phase. There are significant differences between the outcome of these two technology acquisition strategies. In the case of R&D, the technology has been developed in-house and so the company is familiar with it and the information amassed during the project is accessible. In the case of an external transfer of technology, it has to be brought in-house and you have to make sure that all the information you need has been obtained.

While the R&D approach has the advantage that you have access to the information you have amassed during the project, the question arises about the adequacy of this information to deal with unforeseen issues. In technology transfer you can call on the experience of the licensor to help you deal with problems as they should already have amassed the knowledge to deal with them during the development and exploitation of the technology.

Other issues arise if the technology has not been commercialised in the form in which you intend to use it. This could be semi-developed technology from a university or a new application for an existing element of software. Here you may be on your own if things go wrong and you have to make sure that, as far as possible, you have all the information you need to complete the project.

Technology transfer period and deliverables

The agreement should specify the technology transfer period and deliverables. This will vary from one project to another. Examples of how technology might be transferred are listed below:

- Visit of your personnel to train at licensor's premises.
- Visit of licensor's personnel to install and commission equipment and train personnel at your premises.
- Purchase and installation and commissioning of other equipment specified by licensor.
- Supply of documentation in electronic or other form.
- Supply of software, debugging and commissioning.
- Ongoing consultancy or design input from licensor.
- Joint development work at either licensor's or licensee's premises.
- Pilot production by licensor to fill product pipeline for licensee until production is running smoothly.

Regardless of the type of project that may involve many or all of the above methods of technology transfer and start-up assistance, the licensee must ensure that resources are provided to ensure the correct transfer and implementation of the technology.

In terms of what is to be provided by the licensor, the licensee must ensure that the technology transfer period is long enough to transfer all of the information and get the new technology running correctly. In terms of what is to be provided by the licensee, it is essential to be ready to receive the injection of new technology. This can include the following:

- Employing suitably qualified and experienced staff to learn the new technology and operate it.
- Purchasing the necessary production equipment and having it on-site in time.
- Setting up documentation systems to ensure that the new technology is recorded and accessible to staff after the licensor's people have left.
- Giving special attention to anticipating unforeseen problems.

The licensee may also need to take account of IP rights licensed to it and may need to mark products or put notices on documentation concerning licensed rights or warning notices concerning patents or other rights. The licensee also needs to ensure that marketing strategies take account of monopoly positions created by the IP. For example, a standard pricing strategy for launching a patented product is that of market skimming. You enter the market at a relatively high price, build up sales in a controlled manner and then reduce prices later when you are sure you can supply in volumes.

Problems in technology transfer

Problems can arise from a lack of preparedness by the licensor or by the licensee. If the licensor has not transferred its technology before, it may happen that the task of transferring the technology has been underestimated. Frequently, the information required is not written down but has existed in the form of know-how and experience. In some cases the people transferring the technology have never had to train others in the use of it and may not be able to present the information in the correct form or answer difficult questions.

In the case of the licensee, the problems usually arise from the step-change in technology taking place. This may be beyond the technical education of existing staff and it is usually wise to recruit new people with the required education and training.

Many problems do not become evident until the time comes to switch on the equipment. Errors or omissions then come to the fore and it is important to have built up good relations between the staff on both sides so that issues can be resolved without having to question or revisit the technology transfer contract. For example, if quality problems arise in production, this can be due to a failure by the licensor to transfer the necessary skills to the licensee or by a failure of the licensee to understand, document or implement the knowledge transferred. An unprepared licensee may blame the licensor's technology saying that it does not perform as promised. An unprepared licensor may blame the licensee's staff saying that they are not sufficiently competent technically.

Technical assistance

Technical assistance is the help that can be given by the licensor to the licensee after the technology transfer period is over. This should be negotiated as part of the contract and would normally be stated as a consultancy service to be provided by the licensor at the specific request of the licensee. The licensee is usually liable for travel and subsistence expenses of the licensor's staff as in the case of the technology transfer period. A daily rate for the licensor's staff should be agreed in writing as part of the contract.

If the planning and implementation of the technology transfer has been done correctly by both sides, the technology should be operating correctly by the end of the technology transfer period and the technical assistance provisions should not need to be used. Ideally, these provisions are for unforeseen problems down the line or upgrades or modifications that become necessary.

Commercial exploitation strategies

In general, technology is commercially exploited by making and selling a product or providing a service. This method of exploitation is appropriate to manufacturing and trading companies but research organisations and universities do not manufacture or trade. For these bodies, licensing or sale of technology is the only method of commercial exploitation. Different development and protection strategies are required for each type of organisation.

Manufacture and sale or service provision

This type of organisation usually does its R&D in-house or at least in secret so that competitors will not know what is going to emerge. This means that patent applications should be made when all technical problem-solving stages are complete, which may lead to delaying the filing of patent applications somewhat. The technical details are only disclosed when the project is market-ready and patent publications occur later, thus limiting access to the developer's plans by competitors.

Research organisations

Research organisations need to license or sell their technology and they usually do this at an early stage. They tend to patent and disclose early and seek licensees before they have incurred too much expense on development or patenting. For them, carrying the development too far towards commercialisation can be counterproductive as a licensee will have their own views about how the technology is to be commercialised, or how the development is to be completed to meet the market needs they perceive.

Licensing-out

Licensing-out is the activity engaged in by a technology developer. The developer may be a research organisation or college, or it may be a commercial company that has seen the advantages of licensing-out in addition to or as an alternative to manufacture and sale or provision of services. Possible reasons for licensing out are:

- Only available commercial strategy for the organisation.
- Exploiting a distant market.
- Exploiting spin-off technology.
- Non-exclusive licensing of technology to competitors.

The principles described in the foregoing chapters on the evaluation of licences, the setting of royalty rates and negotiation principles apply equally to licensing-out as to licensing-in for technology acquisition.

Open innovation and new business models

Licensing-out is also part of the open innovation strategy discussed in Chapter 1. Increasingly, newer business models are being adopted as companies seek to find new ways to leverage their intellectual assets for maximum advantage. In some cases companies decide not to protect their technology but rely on the downstream benefits of its adoption. In other cases a business model has been adopted that makes products available below cost but captures the revenues from the use of the products.

The concept of open source software has been developed as a reaction to the closed innovation strategies of companies who jealously guard their source code. The developers of the Linux operating system believed that the software community would benefit by allowing others to participate in its development and to share their combined efforts. Intel encourages others to develop products that will increase the demand for its own products. It often publishes its research results thereby making them available and preventing others from patenting them.[1] *Research Disclosure* is a publication that exists for the purpose of making research results available for organisations who do not want to avail of the patent system but want to make sure their own innovations will not be patented by others. It is published in hard copy and as a searchable database and provides a date verification for any disclosure that can be of value in issues concerning prior art.[2] Thus new developments are placed in the public domain thereby destroying their novelty in patent law and making them freely available to all including the originator.

Many products are sold below cost to capture the downstream revenues for their use. Examples are printers where you have to purchase expensive ink cartridges and mobile phones where the network relies on the revenues from call charges. One of the pioneers of this model was Xerox, which had developed plain paper copying technology but found that its copiers would be expensive to purchase. Instead it decided to lease its copiers and to charge only $95 per month. However, if the customer made more than 2000 copies per month they paid 4 cents per copy thereafter. Xerox correctly guessed that once people experienced the convenience of plain paper copiers they would exceed the rather low usage of their coated paper competitors. This strategy succeeded beyond their wildest dreams with Xerox revenues reaching $2.5 billion by 1972.[3]

Product liability

In introducing or marketing new products, whether the technology was acquired by development or licensing, a technology manager needs to be aware of the legal implications of putting a new product on the market. New legislation on consumer protection has placed the responsibility for making sure that products are safe squarely on the shoulders of the product developer who needs to be aware of the issues that come under the heading of product liability.

Product liability legislation has been introduced in Europe following an EU directive in 1985. Prior to this, if a person suffered some injury or loss as a result of some fault in a product, it was necessary to prove negligence on the part of the manufacturer or supplier. Under the legislation, it is no longer necessary to prove negligence. It is only necessary to prove that the injury or loss occurred as a result of the product.

Any company contemplating developing a new product must take account of the potential liabilities that could result from developing a defective product. From a strategic and management point of view, it is important to be aware of the fact that there are certain legal risks involved in developing new products and that it is normal to apply for product liability insurance to cover these risks.

History

The origins of product liability go back to the start of the consumer movement in the USA in the 1930s and in Europe after the war. Mass production and mass marketing had led to a situation where large corporations held great power while their customers as individuals had very little. Products were made and sold without due consideration to the consequences and a number of people were injured and some were killed. Things began to change in the USA after a number of well publicised cases involving defective automobiles. The courts threw out sale contracts that had disclaimers limiting the liability of the manufacturers and the government set up the National Commission on Product Safety. This commission reported that self-regulation by the manufacturers was unenforceable and called for legislation, which resulted in the Consumer Product Safety Act of 1972.

In Europe things developed piecemeal and slowly. Problems were caused by the way legal precedents had been established. Some countries such as the UK and Ireland have a common law system where previous cases provide precedents that effectively change the law. In other countries that have a civil law system this is not the case. For example, in the UK if you sued under the Sale of Goods Act, you could only do so if you were the purchaser. What if you bought an electrical appliance that injured a member of your family? They were not privy to the sale contract and so could not sue. The Sale of Goods Act states that the goods should be fit for the purpose intended, and so it could be established that a defective product was not fit for the purpose intended. This could result in a refund of money but it was still necessary to prove negligence if a person was injured or some consequential loss took place. It was therefore felt that legislation was needed to redress the balance and the concept of 'strict liability' was introduced.

Strict liability

Strict liability is defined as 'liability which does not require the plaintiff to prove that the defendant was negligent and does not exempt the defendant from liability if he proves simply that he was not negligent.' Eventually in a Europe-wide initiative an EU directive was adopted in 1985. This is called the 'Council Directive relating to the approximation of the laws, regulations and administrative procedures of the Member States concerning liability for defective products.' It states: The producer shall be liable for damage caused by a defect in his product.' (Article 1). Thus the producer is strictly liable and cannot hide behind disclaimers or a requirement to prove negligence.

The main provisions are:

- Strict liability.
- Manufacturer ('producer') is liable.
- Seller is liable if manufacturer cannot be identified.
- Refers to industrially manufactured goods.
- Allows for a 'State of the Art' defence.
- Financial limits.

The EU directive defines a 'producer' as the manufacturer or producer of a finished product or of a raw material or component or a person who puts his name or trademark to the product or any importer into the EU or a seller if the producer cannot be identified.

Industrially manufactured goods excludes primary agricultural products. Thus, chickens are excluded but frozen chicken pieces are not. Some element of industrial processing is needed to bring the product within the scope of the legislation.

The 'State of the Art' or 'development risks' defence allows a manufacturer to claim that they are not liable for practices considered safe at the time of development but subsequently found to be unsafe. This is particularly relevant in the case of pharmaceuticals.

Is the developer liable?

If the developer is not the producer, the developer is not subject to strict liability. In other words, if the developer is sued, it would be necessary for the manufacturer or the plaintiff to prove negligence on behalf of the developer. This would be the case whether the development was done under contract or licence. However, if a developer also licensed a trademark, they may be liable as they are representing themselves as being the manufacturer.

What should a manufacturer do?

There are a number of prudent management initiatives that the manufacturer should undertake. These are:

- Carry out an audit of all products to make sure they could not become defective (WHIF or FMEA analysis).
- Install quality management systems (ISO 9000).
- Keep adequate records.
- Carry product liability insurance.

WHIF analysis is 'What Happens IF' and with Failure Modes and Effects Analysis (FMEA) it provides a choice of two similar techniques for trying to predict what might go wrong in the use of a product or how the product could be developed to make it safer. This analysis should be applied to all products, both existing and new.

Other factors to consider are the likely practical implications of court action. While the developer is liable, this does not mean the developer has to pay. The court will consider the degree to which the purchaser or user of the product was to blame. Only then will the court decide if the manufacturer has to pay damages. Factors considered would be:

- The circumstances in which the loss or injury took place.
- Was the product used in accordance with the instructions?

- The age of the product.
- The maintenance or state of repair of the product.

It is worthwhile to look carefully at the instructions and specifications of each product and to make sure that the purpose for which the product is intended is clearly stated and that other uses are not recommended. This should be done assuming that people, especially those with no technical knowledge, will do stupid things. A lady in the US was in the habit of drying her cat by putting it sitting in the oven. When she got a microwave oven she did the same thing and became upset when her cat was cooked from the inside out. She sued the manufacturer and won her case because the instructions said 'can be used just like an ordinary oven'.[4]

Standards issues in new product introduction

This section attempts to answer a number of questions that face those developing or marketing new products. These questions arise from the development of standards, both formal and industry, and it is important to know the management implications of standards in new product introduction. Questions which arise are:

- What is a standard?
- Are standards compulsory?
- How are standards drawn up?
- Do standards hinder innovation?
- If a product meets a standard can you be sued under product liability legislation?
- What legal requirements must manufacturers be aware of?

What is a standard?

A standard is a technical specification agreed between a number of parties. The parties are usually industrial, commercial, research or government organisations, and consumers.

Some standards apply only to very specific products, while others apply to a whole range of products of a general class, or to manufacturing or production processes. For example, BS1363 applies to 13-amp electric plugs, while ISO 9000 applies to quality management in general.

The definition of a standard implies that a product or process should be sufficiently well known and established for there to be a need to write a standard for it. Therefore, there is not necessarily a standard that applies to a new product. New products often require the development of appropriate testing methods before standards can be even considered.

Types of standards

There are many different types of standards to cover:

- Quality.
- Safety.
- Performance.
- Durability.
- Compatibility/connectivity.
- Materials.
- Environmental requirements.
- Products and processes or systems.

Are standards compulsory?

In general, standards are voluntary. Most standards arise from a desire to improve trade. The first standards were for weights and measures without which it would be very difficult to buy and sell goods. Standards were then introduced for guns and bullets so that people in different places could buy guns knowing that they would be able to get bullets for them.

Later it became clear that there was a need to control the sale of certain products for safety reasons (e.g. toys, electrical goods). Governments found it convenient to use standards where these could define specifications of products which were considered safe to use. This led to the use of statutory instruments and directives to specify the permissible technical specifications of goods that may be supplied to the public, or to set minimum technical standards for what may legally be supplied.

The answer is therefore that while standards are voluntary they may be referred to in compulsory regulations associated with national or community laws.

How are standards drawn up?

Standards are drawn up when a need for them arises. The impetus usually comes from industry but in the case of safety standards it often comes from government.

There are a number of bodies that act as co-ordinators for the drafting and issuing of standards. The principal international bodies are: the International Electrotechnical Commission (IEC), and the International Organisation for Standards (ISO). Most countries try to follow the standards co-ordinated and issued by these bodies. The US also participates in these bodies but most US manufactures have not followed the international standards, preferring instead to follow a wide variety of US industry standards issued by about four hundred private standards organisations. However, this situation is changing with more US products being internationally traded and the US has now started to introduce metric units into its electrical standards.

In Europe, CEN (Comité Européen de Normalisation) and CENELEC deal respectively with the harmonisation of non-electrical and electrical standards in the EU. There are also national standards bodies in most countries for example, the British Standards Institute (BSI) and the National Standards Authority of Ireland (NSAI). The drawing up of standards involves considerable consultation and it can take a long time to reach consensus of all the parties. For this reason, standards tend to lag behind the introduction of new technology by some time.

Do standards hinder innovation?

There is no such thing as a standard for a completely new product because a product has to be established in the market before a need for a standard can arise. When Philips developed the audio compact cassette it was unique and it became the standard, but with video cassettes there was a long fight between VHS and Betamax before VHS eventually won out. And the best product doesn't always win. Many computer experts believe that MS-DOS became out of date as an operating system for computers but it survived because it had the resources of IBM and Microsoft behind it and they have established it as an industry standard.

Can you think of a really old product standard that we are stuck with which is clearly not ideal in this day and age? How about the Qwerty keyboard? This was developed so that keys would not jam on a mechanical typewriter. The sequence on the keyboard was designed so that letters often used together did not get in each other's way! Instead of grouping frequently associated letters together, they are as far apart as possible.[5] How many extra miles have typists' fingers had to travel because of this?

Thus standards can hinder innovation, but it is not the fault of the standards. If something becomes an industry standard, it is usually because of market dominance or compatibility issues. These kind of standards are not mandatory, merely a market requirement. Mandatory standards can sometimes make it difficult to introduce new products where there is a safety aspect. A new product may have to conform to standards for its general class. You cannot market any new electrical product for general use in Europe unless it conforms to the Low Voltage Directive.

If your product meets a standard can you be sued under product liability legislation?

Meeting a standard is no guarantee of freedom from being sued if your product causes some loss or injury. The following reasons could apply:

- Your product is defective due to a fault in manufacturing. The design may meet the standard but a particular product may be faulty.
- The standard may not deal with the particular situation. It is important to read a standard carefully to see exactly what it specifies. For example, the standard in question may not be one that deals with issues of safety.
- The standard may not be relevant. US electrical standards do not meet European safety requirements in some cases.

What legal requirements must a manufacturer be aware of?

Product directives of the EU deal with the mandatory requirements for the performance or safety of products in various sectors. The directives use the 'new approach', which is distinct from the 'old approach' where each directive contained a detailed list of all the technical requirements to be assessed by government bodies. In the new approach, direct involvement by such bodies is limited to what is absolutely necessary to ensure compliance with specific essential requirements, and conformance to harmonised standards whose references have been published in the *Official Journal of the European Communities* is deemed to indicate and guarantee such compliance. Information on which directives apply in each sector should be obtained from the British Standards Institute or from other national bodies. They can also advise on conformity and testing procedures to meet the legal requirements.

CE marking

The CE conformity marking placed on products by manufacturers or agents indicates to the public that the products comply with the appropriate directives and ensures acceptability of the products throughout the EU. No Member State may introduce national regulations requiring other conformity procedures.[6]

In order to be able to affix a CE mark to a product a company must carry out a review of the relevant harmonised standards and directives that apply to the product. They then need to carry out a hazard analysis and a risk assessment and must compile a technical file showing the steps that have been taken to ensure that the product is safe. It is wise to get advice from the national standards organisation as some products must be certified by an authorised body.

Enforcement of intellectual property

When a company has successfully introduced its new product to the market and it is selling well, the price of commercial success may be some form of imitation. It is essential to have a method of identifying infringements and a policy for enforcement of rights.

All marketing and technical staff should be instructed to be on the lookout for infringements of the company's rights. This requires a written document for staff setting out what the company's IP is and indicating what would constitute an infringement. A close watch on competitors, their new product offerings and their patent applications will pay dividends.

To get evidence of infringement it is usually necessary to buy a copy of the infringing product to prove that it was being offered for sale and to enable a court to see if the product infringed. Getting evidence of infringement of process IP can be more difficult and it may be necessary to enter a competitor's premises to do this.

As stated in Chapter 7, a company should decide on how it intends to defend its rights and as this can be costly, many companies take out IP insurance to cover the cost of losing a legal action. If you cannot afford the cost of going to court, you have to rely on bluff and threats of legal action to enforce your rights. In issuing threats, your bluff may be called so this is a dangerous strategy.

The first step in enforcement of rights is to draw the attention of the alleged infringer to the existence of your rights. This should be done without making any allegations or issuing threats. The next step, after taking legal advice as to whether there is a likely infringement, is to issue a 'cease and desist' letter. This, as its name implies, is a warning and a threat of legal action and is usually issued by a patent attorney. Such threats should not be issued lightly as the alleged infringer can sue for groundless threats and could get damages if it was found that there was no infringement.

Owners of IP rights can go to court for injunctions to force an infringer to stop selling a product until a court hearing is held to decide the matter. They can also go to court to obtain orders called Anton Piller orders, which enable a holder of IP rights to enter an alleged infringer's premises.

Notes

1. Chesbrough, H.W. (2003) *Open Innovation: The New Imperative for Creating and Profiting from Technology*. Harvard Business School Press, p. 172.
2. See *http://www.researchdisclosure.com* (accessed on 18/04/06).

3. Chesbrough op.cit. pp. 71–74.
4. The author has read a newspaper report on this in the past and believes it to be true. It has now become an urban legend and has been retold with variations. See *http://www.snopes.com/horrors/techno/micropet.htm* (accessed on 12/04/06).
5. See *http://home.earthlink.net/~dcrehr/whyqwert.html* (accessed on 31/05/06.
6. See *http://europa.eu.int/comm/enterprise/newapproach/legislation/guide/legislation.htm* (accessed on 12/04/06).

Appendix 1
On-line information sources

Technical journals

http://pubs.acs.org/journals/aoc/aoc_search.html	American Chemical Journals
http://bubl.ac.uk	UK journals
http://www.tandf.co.uk/journals/alphalist.asp	UK Journals
http://www.eevl.ac.uk/eese	Engineering & Computing ejournals
http://www.asme.org/pubs/journals	American Engineering
http://eels.lub.lu.se	Engineering E-Library
http://www.ili.co.uk/en	Standards
http://www.cenorm.be	European Standards
http://www.dialog.com	Dialogweb is friendlier
http://www.dialogweb.com	Dialogweb
http://www.ncbi.nlm.nih.gov/PubMed	Medical

Market information

http://www.profound.co.uk	Can be accessed through Dialog
http://www.thomasregister.com	Trade Directory of the USA
http://www.europages.com	European Business directory
http://search2.ft.com	Financial Times
http://www.eiu.com	Economist Intelligence Unit
http://www.kompass.com	Kompass worldwide trade directories

http://www.hoovers.com	US company information
http://www.dnb.com	Dunn & Bradstreet, financial data

Sources of technology or licences

http://www.autm.net	Association of University Technology Managers
http://www.yet2.com	Licensor/licensee matchmaking
http://pharmalicensing.com	Pharmaceutical industry licences
http://irc.cordis.lu	Innovation Relay Centres
http://www.les.org	Licensing Executives Society

Patent databases

The following databases are useful in searching for information on patents:

http://ep.espacenet.com
Free database of European Patent Office. An excellent resource for patent searching.

http://www.espacenet.com/access/index.en.htm
Gives access to all patent offices in Europe

http://www.uspto.gov
The US Patent and Trademark Office. A very informative free website for all US intellectual property matters.

http://www.delphion.com
A subscription database, reasonably priced.

http://www.dialog.com
Dialog subscription database gives access to World Patent Index.

http://www.dialogweb.com
Gives access to World Patent Index.

http://www.questel.orbit.com/index.htm
Subscription database and information services for intellectual property.

http://www.wipo.int
World Intellectual Property Organisation. Information on all aspects of intellectual property and especially international issues, polices, domain names.

http://www.wipo.int/classifications/en/index.html
Direct access to WIPO international patent classifications.

http://www.patentcafe.com
Advanced patent searching and management services.

Tracking the progress of patent applications

http://www.epoline.org/portal/public
For European patent applications. Free service but you must register to get access.

http://portal.uspto.gov/external/portal/pair
For US applications.

Training in patent searching

http://patentinfo.european-patent-office.org/learning

Index

Made in the USA
Monee, IL
13 September 2026

62975587R00133